Allan Næs Gjerding

Technical Innovation and Organisational Change

The Innovation Design Dilemma Revisited

Industrial Development Research Series

Department of Business Studies, Aalborg University

Technical Innovation and Organisational Change

The Innovation Design Dilemma Revisited

First published in 1996

ISBN   87-7307-526-4

©   Allan Næs Gjerding, 1996

Publisher   Department of Business Studies, Aalborg University

Distribution Aalborg University Press
              Badehusvej 16
              DK-9000 Aalborg
              Tel.: +45 98 13 09 15
              Fax: +45 98 13 49 15

Print   Thy Bogtryk & Offset

Cover   Johannes Andersen

Illustration Poul Esting/Stormy Blues

Contents

Preface 5

1.   *A Structural-Functionalist Approach to the Technological Imperative* 11

   1.1 The South Essex experience, 11
   1.2. Proposition and scope of the present study, 15
   1.3. The basic approach and some central concepts, 18
   1.4. The compatibility of the organisation theories, 23
   1.5. Overview and summary of chapters 2-9, 28

2.   *The Flexibility-Stability Dilemma* 43

   2.1. A simple model of intraorganisational dynamics, 43
   2.2. The classic flexibility-stability approach, 48
   2.3. Some critical remarks, 58

3.   *A Behavioural Theory of Organisational Action* 63

   3.1. The apperance of the behavioural theory of the firm, 63
   3.2. Satisficing behaviour and bounded rationality, 64
   3.3. Restating the flexibility-stability approach, 70

4.   *Contingent Decision Making and Problem Solving* 83

   4.1. Integration and differentiation, 83
   4.2. Combining the logic of rational and natural system models, 88
   4.3. Implications to the flexibility-stability dilemma, 96

5.   *Two Cases of Organisational Innovation* 105

   5.1. An action research project, 105
   5.2. Problem: Resistance to organisational change, 109
   5.3. Solution: Visibility at three levels, 114
   5.4. Evidence on the flexibility-stability dilemma, 117

6.   *Learning Cycles and Some Concepts* 123

   6.1. The incompleteness of learning cycles, 123
   6.2. The neo-behavioural approach, 127
   6.3. An evolutionary approach, 132
   6.4. Learning as an interactive process, 145
   6.5. Intersections of neo-behavioural
         and evolutionary concepts of learning, 150

7.   *Paradigms and Activities in the Innovation Process* 153

      7.1. Innovation as a process of feedback, 153
      7.2. Patterns of innovation, 163
      7.3. Technological paradigms and techno-economic changes, 175

8.   *Implementation of High Technology* 185

      8.1. Defining advanced manufacturing technology, 185
      8.2. Organisational learning through creative tensions, 192
      8.3. Some Danish evidence on the flexibility-stability dilemma, 203
      8.4. Looking ahead: Some final remarks
          on skills and organisational principles, 213

9.   *New Management Principles and Quality Control* 217

      9.1. Flexibility-stability dilemmas in an ideal-type Japanese setting, 217
      9.2. The opportunities provided by quality management, 225

10.   *Concluding remarks* 235

      10.1. The innovation design dilemma revisited, 235
      10.2. Condensed answers to broad questions, 236

References 243

Preface

Purpose and general analytical approach

The present monograph is yet another study on how organisations learn and innovate. This topic is far from new, and a large number of important studies have already been made by scholars superior to me in scientific skills and seniority. A young(ish) researcher like myself may find the topic interesting, but at the same time have little hope to supply something new in the light of what has already been said. Why, then, embark on a monograph like the present one? Well, writing a Ph.D. dissertation is, of course, an excellent excuse. However, a couple of scientific points of interest have played an important part in the decision, as well. *First,* being an economist I have wondered about what goes on inside the very firms that enter the economic analysis as more or less anonymous entities or aggregates. To an economist, organisational behaviour appears as a black-box described by highly theoretical behavioural concepts which may be very difficult to recognise in real-life settings. The only way to overcome this unfortunate analytical point of departure is to embark on a study of what really takes place inside these entities. *Second,* I have for a number of years been occupied with the study of technical innovation, but have increasingly found it difficult to explain technical innovation without reference to organisational behaviour. Organisational behaviour have, of course, traditionally been the topic of organisation theory, while technical innovation has remained a theme explored by primarily engineers and innovation economists. At some point of time, any innovation economist like myself may feel the need to trespass on this analytical division of labour.

The present study make an attempt to bridge the gap between organisation theory and innovation economics. It is my hope to stimulate the interest of students in the fields of organisation theory and innovation economics towards trespassing the analytic division of labour between their fields of study. The construction of a bridge is far from trivial, and the present study is but a modest attempt to delve into analytical fields that rightfully may belong to scholars other than me.

The study is undertaken within the tradition of innovation economics, and it takes as its point of departure a specific analytical perspective on organisational behaviour, i.e. the idea of a flexibility-stability dilemma (Zaltman, Duncan & Holbek, 1973), later rephrased as an innovation design dilemma (Holbek, 1988) where the notion of *design* refers to organisation design. Put simply, this approach suggests that organisations find themselves confronted with a dilemma in the sense that they have to sustain stability of coordination in order to achieve a certain level of performance and at the same time be flexible in order

stability vs. flexibility

5

organisational vs. technical

to adapt to changing circumstances. Stability of coordination is described in terms of the *persistence* of the nature of organisational activities, while flexibility is described in terms of the organisational ability to initiate and implement organisational and technical *innovation*. Organisational innovation refers to a change of the organisational configuration, while technical innovation refers to a change of the organisation's process and/or product technology. Innovation is a necessity for organisations if they are to survive on a long run basis in a competitive environment, but innovations can only contribute effectively to organisational survival if the change of activities implied by the innovation is aligned with the coordination of new and persisting activities. Thus innovation represents a change of current activities which transforms into the stability of future activities. The present study focusses primarily on a single aspect of the flexibility-stability issue: The transition from the initiation to the implementation of an innovation. The idea is that while the initiation requires the organisation to adopt a flexible approach to its configuration, process and product technology, the implementation requires that organisational activities become routinised. The transition is far from simple and rely on contradictory forces; thus the notion of a flexibility-stability *dilemma*. The way in which this dilemma is resolved depends on the extent to which the processes of learning entailed in the transition contributes to the *adaptability* of the organisation. *learning influence*

The theoretical foundation for applying this approach within the field of innovation economics is evolutionary theory in a general sense, i.e. a theory that

> proposes that the variable or system in question is subject to somewhat random variation or perturbation, and also that there are mechanisms that systematically winnow on that variation. Much of the predictive or explanatory power of that theory rests with its specification of the systematic selection forces. It is presumed that there are strong inertial tendencies preserving what has survived the selection process. However, in many cases there are also forces that continue to introduce new variety, which is further grist for the selection mill.
>
> (Nelson, 1995, p.54)

Basically, the evolutionary set-up implied by the present study operates with the following analytical properties which come in pairs:

(1) There are two sources of variation. *First*, innovation may occur as the outcome of deliberately organised activities in the sense that organisational and technical innovation are implied by the investment projects in process and/or product technology which an organisation undertakes. Furthermore, organisations often organise in order to be able to innovate, often by devoting some subunits to innovation as for instance departments for R&D and human resource development. However, although these activities may be deliberately organised, the actual outcome may not have been anticipated in advance.

organising innovation

Second, innovation may occur as the outcome of the perceptions of the organisation members. Organisation members frequently find themselves in circumstances of discrepancy between what they do and what they think they should do, or what they are actually doing and what they thought they were doing. To the extent that the perceptions of organisation members result in changes of individual behaviour which transforms into changes of organisational behaviour, innovation occurs. However, this transformation may be broken at a number of instances, and the outcome may be far from intended. *selection*

(2) There are two forces of selection. *First*, the organisation operates at a market, i.e. a competitive environment comprised by costumers and competitors. The choices and activities undertaken by these costumers and competitors change the distribution of market shares and thus determine the extent to which the organisation achieves a satisfactory level of performance. Especially, new costumer demands and the introduction of innovations by competitors tends to render the organisation's existing services obsolete. The organisations which are able to introduce innovations which meet an effective *Roles* demand and/or adapt to changing market configurations are those which survive in the *of* long run. *Second*, the organisational configuration defines a set of formal and informal *selection* organisational roles that set boundaries for what the organisation members (can) do and percieve. There exists an intraorganisational division of labour and power which tends to ✗ constrain the rate of organisational and technical change. Applying the idea that organisation members perform certain tasks within the organisation in order to secure that the organisation performs certain tasks at the market, the present study employs the notion of an internal and external task environment which both acts as selective pressures on organisational activities. *Division of labor & power*

(3) There are two sources of inertial tendencies towards preserving what has survived the selection process. *First*, to the extent that organisations have successfully adapted to the external task environment in the past, they persist at the expense of other organisations less adaptable or fortunate. In consequence, the external selection environment tends towards a state of complacency, and the organisations in question remain fit until the sources of variation gain renewed momentum. *Second*, and similarily, to the extent that the nature of existing activities within the organisation yield satisfactory results, the organisation members come to believe that the way in which things are done is effective. In consequence, the internal selection environment tends towards complacency, and the division of labour and power remains stable until the sources of variation gain renewed momentum.

Three aspects of what has been said so far may be noted. *First*, the mechanisms of variation, selection and preservation are interactive. While the forces of selection tend to create tendencies towards preservation which set limits to the forces of variation, the

- *source of variation gains momentum,*
- *it is assumed that the way in which things are done is effective*

7

forces of variation contributes to the creation of new selection pressures that challenges the tendencies of preservation. *Second*, although the organisation may initiate change, the main line of causation runs from the external task environment to the internal task environment. That is, the organisation can only affect the external task environment to a very small degree, and thus the present study regards the organisation as mainly an adaptive unit. However, adaptation does not only occur as a response to external forces of variation, but may also occur as a response to internal forces of variation. Internal forces of variation may occur on the basis of perceptions of future challenges initiated within the external task environment. In consequence, adaptation may be proactive as well as reactive. *Third*, the present study assumes that there exists a satisfactory level of organisational performance, which sets limits to how much effort the organisation may devote to adaptation. Devoting resources stops when the organisational performance meets the level of satisfaction whether or not a higher level might be reached. *getting by*

Within this framework, chapter 1 decribes in detail the specific analytical questions and discusses the methodological approach of the present study, while chapter 2 presents an account of the flexibility-stability issue. There is no need for the preface to repeat the arguments. However, it is necessary at this introductory point of time to comment briefly on an analytical perspective to which the present study devotes scarce attention. Within organisation theory and innovation economics, the notion of interorganisational dynamics has come to play an important part. Increasingly, organisational behaviour is seen as an interorganisational phenomenon, and innovation is described in terms of externalities ranging from the utilisation of general scientific input to the close cooperation between firms and business managers. However, although I recognise the importance of such interorganisational phenomena and have, to some extent, studied them myself, the present study abstracts in general from interorganisational solutions to the flexibility-stability, but for a few exceptions in the chapters to come. The reason for this is that organisational behaviour, to an important extent, is determined by intraorganisational contradictory forces which must be studied in their own right if you are to understand what goes on inside organisations. The present study attempts to do so within the framework described in chapter 2. However, as evidenced by especially chapters 4 and 8-9 this does not imply that you regard the organisation as a self-contained entity sealed off from the outer world.

Following this basic approach, a triple purpose is pursued: First, certain classic organisation theories which have become doctrines are studied and used for reinterpreting the flexibility-stability approach; second, some propositions is made on what innovation economics may learn from these doctrines; third, the observations made so far is used as a contribution to the explanation of current organisational changes associated with the diffusion and implementation of advanced manufacturing technology. Throughout the study, the focus is on manufacturing firms; however much of the discussion takes place at the level of general theorising applicable to administrative and service organisations as well.

The way in which the purpose of this study is pursued reflects, of course, my educational and scientific background. I started out with a BA in business administration, then got a MA in economics, and finally worked for six 'years as a researcher in the zone between macro- and innovation economics. The study is biased towards how I think as an economist and represents an elaboration of the main body of my work as a member of the IKE group at Aalborg University during the period of 1987-93. This research may be divided into three lines of occupation, namely the study of

(1)   a Danish productivity dilemma during the economic boom of the mid-eighties;
(2)   how productivity is modelled in the main Danish econometric models; and
(3)   how organisational performance may be explained by the interplay between technical and organisational innovation.

The present study relies on (1) and (3), while (2) is only partially represented as a source of inspiration to the way in which I think about cause-effect chains. The most important part of my work in line (1) was done within a collective of researchers and is documented in Gjerding et al. (1988, 1990, 1990a), Gjerding & Lundvall (1990), Gjerding (1991) and Madsen et al. (1993); line (2) has been documented in Gjerding & Kallehauge (1990) and Gjerding (1991a); and line 3 is documented in Gjerding (1990, 1992, 1992a, 1994, 1994a), Gjerding & Madsen (1991), Gjerding & Lundvall (1992) and Gjerding & Lauridsen (1995).

Acknowledgements

During the years, it has been stimulating, intellectually and socially, to work within the IKE group. Although the degree of scientific integration between the various ongoing projects tend to vary in a cyclical manner, the group remains a coherent collective despite the tiresome and continous problems of financing the occupations of the junior members.

However, despite the inspiring context, I did not succeed in completing my Ph.D. thesis. Presumably, I was too occupied with too much at the same time. At the most critical point of time, the untimely death of my mother and my brother within five month deprived me of the psychological energy necessary to complete a Ph.D. thesis, and constant worries about my occupational security at the Faculty of Social Sciences were far from stimulating. Finally, in June 1993 I left Aalborg University and took up a position at Nordjyllands Trafikselskab (the North Jutland company for public transportation), where I have been working in the fields of accounting & finance, planning and quality control.

Fortunately, during the Autumn of 1994, a grant from Aalborg University allowed me to take a two months leave in order to complete my thesis. During this period of time, final versions of chapters 1-5 and draft versions of chapters 6-9 were prepared. I am grateful to Nordjyllands Trafikselskab for permitting me a two months leave, to the Faculty of Social Sciences for financing the leave and to the Department of Business Studies for the office facilities provided. I would like to thank my dear colleagues *Jens Mogensen* and *Henrik Dalsgaard*, Nordjyllands Trafikselskab, who had to put up with my tasks as well as their own while I was away.

Within the IKE group, I am especially endebted to *Bengt-Åke Lundvall* and *Poul Thøis Madsen* with whom I have had the opportunity to work in a team-like fashion; to *Birgitte Gregersen* who commented on earlier drafts of this study; and most of all to *Björn Johnson* who has been an extremely supporting supervisor throughout the entire process. *Dorte Køster* provided me, as always, with efficient secretarial assistance.

Without the IKE group, I would like to thank *Reinhard Lund* and *Allan Christensen* for comments on organisation theory literature, *Michael Fast* for a number of interesting discussions, and *Otto Bredsten* and *Brita Lauridsen* with whom I had the opportunity to work at the project described in chapter 5.

The thesis has been completed during April-December 1995 alongside my full-time employment at Nordjyllands Trafikselskab. I am forever endebted to my wife, *Ann Birthe Jensen*, and to my daughter, *Signe*, for their patience during this prolonged process of long working hours and absence.

Nørresundby, April 1996

Allan Næs Gjerding.

Chapter 1

A Structural-Functionalist Approach to the Technological Imperative

1.1. The South Essex experience

The present study is inspired by the following tale, which now have become widely accepted among theorists in the field of organisation theory, industrial economics, and the economics of innovation:

Once upon a time, Woodward (1958, 1965) suggested that the way in which a manufacturing firm organised its productive activities was highly sensitive to the type of production techniques applied. According to the research undertaken by her and her associates in South Essex of 100 manufacturing firms during the period of 1954-58, type of production technique and organisational characteristics were correlated in a way which indicated that the complexity of the production technique was matched by the complexity of the organisational configuration. Especially, there seemed to be a direct relationship between production technique and

> the length of the line of command; the span of control of the chief executive; the percentage of total turnover allocated to the payment of wages and salaries, and the ratios of managers to total personnel, of clerical and administrative staff to manual workers, of direct to indirect labour, and of graduate to non-graduate supervision in production departments.
> (Woodward, 1965, p.51)

The complexity of production technique was described in terms of a continuum of three main production types, i.e. unit, mass, and process production, disaggregated into nine sub-types[1], where the notion of *technical complexity* referred to the fact that "it becomes increasingly possible to exercise control over manufacturing operations" (ibid., p.40) as we move along the continuum from unit to process production. Analysing the organisation of the production departments in terms of the span of control, Woodward (1965) noticed that there were similarities between the extreme positions at the continuum where organic management principles seemed to predominate, while mechanistic principles characterised by the line-staff type of organisation were predominant in the middle range (ibid., p.64).[2] However, the type of organic management principles at the extreme varied according to

---

1. And a residual of combinations of the three types, disaggregated into two sub-types.

2. The notion of organic and mechanistic principles appeared in the work by Burns & Stalker (1961) which are presented in chapter 4. Mechanistic principles refer to the Weberian bureaucracy, while organic principles refer to the antithesis of bureaucracy. See figure 17 for further details.

Mechanistic = bureaucracy
Organic = opposite

the number of management levels, which was higher at the process production end of the continuum, and the number of direct workers to indirect workers, which was higher at the unit production end of the continuum. Figure 1 depicts these findings.[3]

The main observation to be derived from figure 1 is that the tendency towards bureaucratic forms of organisation is largest at the middle range of the continuum and smallest at the extreme ends of the continuum. While tasks and responsibilities were clearly defined in the middle range, the extreme ends exhibited a much larger degree of delegation of authority and decision making. At the unit end of the continuum, management responsibilities of first line supervisors depended on craftsmanship and experience, and the number of specialists to be found in the production department was rather small. At the process end of the continuum, the number of specialists was large, and although the number of levels of management was relatively high, it was, in practice, difficult to distinguish between managers and subordinates. Furthermore, the organisational configuration seemed to be changing, since there was a tendency towards functional or line principles instead of line-staff principles of organisation.

buroucracy

Figure 1. *The relationship between technical complexity and bureaucracy*

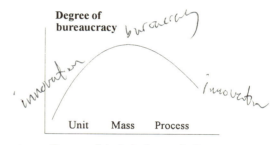

Degree of technical complexity

Source: Woodward (1965, pp.50-67)

Conflict b/w departments

Due to the more organic nature of the extreme ends of the continuum, organisational tensions in the form of interdepartmental conflicts seemed to occur less frequently than in the middle range of the continuum. According to Woodward (1965, pp.125-53), there was a larger degree of integration between the social and the technical systems of the firm at the extremes than in the middle range. In mass production, the organisational configuration was characterised by a high degree of departmental independence where the end results of one department did not depend on the end results of other departments. There was a lack of close interdepartmental cooperation which seemed to "encourage sectional interests and exaggerate

little cooperation

3. The basic idea of figure 1 is adapted from Bruzelius & Skärvad (1979, p.103).

departmental loyalties" (ibid., p.137), and channels of communication were sparse. The division of responsibilities between the departments of research & development and production was blurred, and there existed some hostility between R&D and production personnel. Furthermore, the perceptions of the personnel of the two departments differed somewhat, since the R&D personnel tended to be more oriented towards long-term objectives, less status-oriented among themselves, and less focused on profitability and cost reduction than the production personnel (ibid., p.140). Potentials for conflict were also present in the relationship between the marketing and production departments, since the marketing personnel was more oriented towards costumer needs than the manufacturability of products (ibid., p.149-50). *ibid?* *liason devices*

The way in which these potentials for conflict were bridged was primarily in terms of committees, task forces, or special product development departments as liason devices for research and production. These arrangements were especially important in the case of the production of a new product, where a tight relationship between the R&D and production departments becomes crucial. However, in the course of normal activities, i.e. productive activities associated with existing products and the improvement of existing products, there seemed to be a trade-off between on the one hand the involvement of the R&D personnel in production and on the other hand their focus of attention on research and development. In sum, Woodward and associates argued in favour of a seperation between R&D and production in cases where new products were not involved, "providing adequate channels of communication can be established", since "in normal circumstances the co-ordination between research and production is merely a matter of exchanging information" (ibid., p.141). *new products vs. old not*

While most of the findings presented in Woodward (1958, 1965) have been widely accepted and affirmed through subsequent studies[4], they have also been subjected to some criticism (cf. Tayeb, 1988, pp.12-13). *First*, the idea that there exists a match between the production technique applied and the organisational configuration, in terms of what has been paraphrased a *technological imperative*, presupposes that the production technique of a firm may be described uniformly. This is not always the case, since several types of technical complexity may be found within firms (Khandwalla, 1974; Tayeb, 1988). *Second*, the organisational configuration is influenced by factors other than technical complexity[5], notably the principles of control which reflect the nature of the task

---

4. Clegg (1990, p.56) argues: "The findings now seem remarkably commonsensical, although prior to Woodward's work they had not been grasped. It is clearly a measure of her achievement that they now seem so obvious".

5. As later recognised and also emphasised by Woodward (1970).

environment (Thompson, 1967). *Third*, the Woodward findings seem to be highly influenced by the size of the firms in her sample. In a subsequent study[6], Hickson, Pugh and Pheysey (1969) reported themselves unable to verify Woodward's findings unless the aspect of size was taken into consideration. However, this might be explained by the fact that the impact of production technique on the organisational configuration tends to vary positively with the share of production operations in total activities. That is, the smaller the firm, the larger the impact of its operating core, and the larger the firm, the larger the relative independence of the administrative and hierarchial structure from the operating core.

Thus, according to the Aston findings, the degree to which production technique affects the organisational configuration depended on the importance of size in terms of interdepartmental dependencies. However, in his study on 79 manufacturing firms[7], Khandwalla (1974) was not able to confirm neither the findings of Woodward (1965) nor the findings of Hickson et al. (1969). Reviewing a number of studies in relation to the South Essex and Aston findings, he argued that there "is no uanimity that technology affects organizational structure", and that the "impact of technology on organizational structure, if any, is likely to be quite selective" (Khandwalla, 1974, p.77). On the other hand, some evidence in support of a soft version of the technological imperative were obtained. Dividing the sample according to profitability into a high-profit and a low-profit sub-sample which were rather similar according to size and industry distribution, Khandwalla (1974, pp.90-96) found that the relationship between production technique and the organisational configuration was positive in the high-profit group and negative in the low-profit group. Thus, there seemed to be a technology match which affected the economic performance of the firm:

---

6. The study reported by Hickson et al. (1969), which is part of the Aston studies, relies on a sample from the Birmingham area where the size of firms was somewhat larger than in the South Essex sample: "The sizes of the organizations in south-east Essex ranged down to 100 employees, whereas the Birmingham sample had a minimum size of 250; and there were proportionately fewer multithousand units in Essex than in Birmingham" (Hickson et al., 1969, p.394). The Birmingham sample contained 52 organisations of which 31 were manufacturing firms, and the following refers to the investigation of these manufacturing firms.

7. In order to assess the type of production technique of each firm, the respondents were asked to attach ratings to a scale of production techniques comparable to the Woodward scale, and subsequently weights were assigned to each point of the scale in order to measure the degree of mass output orientation of the firm. The relationship between production technique and organisational configuration was measured in terms of the delegation of authority (decentralisation) and the use of sophisticated control procedures.

It was argued earlier that mass-output orientation of technology, by professionalizing production-related decision making, may lead to greater functional specialization and therefore to greater decentralization and greater use of controls. This line of reasoning recieves some support for the high-profit group. If true, it implies that the low-profit firms make a maladaptive response to technology or, at any rate, one that is rather different from that made by the profitable firms, such as providing coordination by centralizing decision making in the face of technology-induced functional specialization, rather than through the use of sophisticated controls.

(Khandwalla, 1974, pp.94-95)

technology match *

During the recent years, the technology match hypothesis has found its way into the field of innovation economics. This has taken place gradually and with some difficulty, since innovation economics *qua natura* focusses on the technical rather than the organisational aspects of the technology match. The interest of innovation economics in the technology match has increasingly been stimulated by the advent of challenges to the dominant Western-style principles of technical innovation and organisational change, notably the increasing relative economic performance of Japanese firms *vis-à-vis* their Western adversaries[8], and the issue of a second industrial divide (Piore & Sabel, 1984) which imply that firms may be flexible, even though they specialise (Miles & Snow, 1986; Clark & Starkey, 1988). However, innovation economics is still in "search for a useful theory" (Clark & Starkey, 1988; Clark & Staunton, 1989) regarding the interplay between technical innovation and organisational change. *technical + organizational*

The present study is a very modest attempt to indicate some lines of thought along which this search for a useful theory may proceed. It is based on the assumption that the technology match hypothesis is valid in the sense that firms exhibit a soft variant of the technological imperative. This is not to say that a strong relationship between production techniques and organisational configuration may be found in every firm, but it implies that the relationship tends to become stronger in those firms which are most successful in terms of economic performance. *relationship stronger in more successful firms*

1.2. Proposition and scope of the present study

The reason why the present study is inspired by the idea of a technological imperative is the following: During the mid-eighties, the Danish economy experienced a productive decline in the midst of economic prosperity. Despite high growth rates of production, investments, employment and profits, productivity growth deteriorated, even at negative growth rates in the manufacturing sector during the years of 1984-86. As argued

---

8. For instance, Freeman (1987, 1988), Clark & Staunton (1989), Dertouzos et al. (1989), Whittaker (1990), Roos (1991), Gjerding (1992).

15

elsewhere (Gjerding et al., 1990, 1990a), there were several reasons for this, one of which was the apparant difficulty of accommodating labour skills and work organisation to the implemention of advanced manufacturing technology (Gjerding & Lundvall, 1990), which diffused rapidly in the Danish economy during the period in question (Industri- og Handelsstyrelsen, 1988; Kallehauge, 1990). As appears from chapter 8, the same phenomenon has been reported elsewhere, e.g. in the US (Dertouzos, Lester & Solow, 1989), the UK and France (Zairi, 1992).

The apparant paradox of declining productivity in the midst of the growth of virtually everything else raises the issue of the relationship between technical and organisational innovation. The present study discusses this issue from the point of view of a simple idea which is pursued by the discussion of three research questions that reflect a triple purpose.

The simple idea

On the one hand, organisations must remain stable in order to achieve a certain level of performance. On the other hand, organisations must be flexible in order to adapt to changing circumstances. Stability is achieved through coordination which ensures that the nature of organisational activities tend to persist over time. Flexibility is achieved through organisational and technical innovation. Organisational innovation refers to a change of the configuration of the organisation, while technical innovation refers to a change of process and product technology.

Innovation confronts the organisation with a disturbance of stability, since innovation represents a change in the way activities are carried out. Thus, the present study argues that there may be a dilemma entailed in innovation, which relates to the transition from the initiation to the implementation of innovation. At the initiation stage, the organisation must gather and process new knowledge associated with the generation and evaluation of proposals for innovation. At the implementation stage, the organisation must institute rules of procedures. In consequence, while the initiation requires that the organisation adopts a flexible approach to its configuration, process and product technology, the implementation requires that organisational activities become routinised. The transition is far from simple and may rely on contradictory forces. The present study employs the notion of a flexibility-stability *dilemma*, which is described in some detail in chapter 2, and argues that the way in which this dilemma is resolved depends on the learning processes entailed in the transition.

The notion of a flexibility-stability dilemma rests on the assumption that innovation is initiated because the organisation members percieve a performance gap, defined as a discrepancy between the goals pursued and the goals attained, or between what the organisation members are doing and what they think they should be doing, as Zaltman,

Duncan & Holbek (1973) put it. Chapter 2 has more to say about the occurrence of performance gaps, and chapter 3 explains the occurrence of performance gaps in terms of deviations between levels of aspiration and levels of achievements.

The three research questions

The three research questions are the following:

theory of the firm & organization

(1)   *How can the flexibility-stability dilemma be described in terms of the classic* action *behavioural theory of the firm and contingency theory of organisational action?*

The flexibility-stability dilemma is presented in the form of the theories of organisational change by Hage & Aiken (1970) and Zaltman, Duncan & Holbek (1973). The propositions of these theories, which are coined the classic flexibility-stability approach, are elaborated in terms of three bodies of organisation theory: The behavioural theory of the firm (Simon, 1957; March & Simon, 1958; Cyert & March, 1963); the contingency theory of organisational action (Burns & Stalker, 1961; Lawrence & Lorsch, 1967; Thompson, 1967; Galbraith, 1977); and the neo-behavioural theory (Cohen, March & Olsen, 1976; March & Olsen, 1976, 1976a). In conclusion, an elaborated version of the flexibility-stability dilemma is described.

(2)   *How can this description be interpreted from the point of view of the economics of innovation?*   economics of innovation

The economics of innovation is represented by studies on three main topics: The change of organisational routines (Arrow, 1962; Rosenberg, 1982; Nelson & Winter, 1982; Lundvall, 1985; Johnson, 1992); the innovation process (Freeman, 1982; Cooper, 1983; Pavitt, 1984; Saren, 1984; Kline & Rosenberg, 1986; Christensen, J.F., 1992); and the existence of technological paradigms (Nelson & Winter, 1977; Dosi, 1982; Perez, 1983; Sahal, 1985; Freeman & Perez, 1988).

(3)   *How may this description contribute to the explanation of the contemporary organisational changes which are most frequently associated with the diffusion and implementation of microelectronics and information technology?*

The flexibility-stability approach is confronted with three types of empirical evidence: Evidence on the diffusion and implementation of microelectronics and information technology in the Danish industry during the 1980s (Gjerding & Lundvall, 1990; Gjerding et al., 1990, 1990a; Kristensen, 1992; Nyholm et al., 1994); case study of organisational change in two Danish manufacturing firms during 1993-94 (Gjerding & Lauridsen, 1995); and the debate on the idealtype Japanese principles of management and quality control versus the principles of idealtype Fordism (Deeming, 1982, 1986; Freeman, 1987, 1988; Dertouzos, Lester & Solow, 1989; Aoki, 1990; Shimada, 1991; Gjerding, 1992, 1994, 1994a).

Fordism + JMS

The triple purpose

These three questions reflect that the present study has a triple purpose: First, it is a study of certain classic organisation theories which have become doctrines; second, it is a proposition of what the economics of innovation may learn from these doctrines; and third it is a discussion of how these doctrines may contribute to the explanation of current organisational changes associated with the diffusion and implementation of advanced manufacturing technology.

The third purpose relates to an issue which is the subject for a fierce contemporary debate. Some theorists have argued in favour of a new type of organisation theory (Clark & Starkey, 1988; Clark & Staunton, 1989), even a postmodern one (Clegg, 1990; Gergen, 1992), from the point of view that we are entering a post-Fordist era (Piore & Sabel, 1984). However, the present study adopts a more cautious point of view like Whitaker (1992), but still argues that the implementation of advanced manufacturing technology makes the focus on consultative industrial relations, learning processes and human resource development more important (Haywood & Bessant, 1987; Adler, 1990; Gjerding et al., 1990; Leonard-Barton, 1990; Tyre, 1991; Zairi, 1992) than suggested by the classic organisation theories.

Structural functional

1.3. The basic approach and some central concepts

The overall approach of the present study is evolutionary, as described previously in the preface. During the analysis, this general evolutionary approach is operationalised in terms of a structural-functionalist approach in the sense that the discussions try to explain "the existence and continuity of structural elements, in terms of the contribution to the rest of the system" (Donaldson, 1985, p.28). Traditionally this means, in the study of organisations, that the constituent parts of the organisation are held together because they contribute to the persistence of the system and at the same time recieve the resources necessary to sustain this contribution. In general, the structural-functionalist paradigm may be described as follows:

> Structural-functional analysis has at its core a model of a social system which, when in equilibrium, has sufficient interchanges between the sub-systems to sustain each of them, and thereby maintain the social system. Equilibrium is thus in a state which is not devoid of movement, for the flows move between sub-systems, but is one in which the overall social

18

system in its environment is stationary. Change takes the form of internal adjustment to reach equilibrium. This is normally concieved of as caused by external changes in the environment to which the social system must adapt. This usually entails adjustment of one or more sub-systems, leading, in turn, to adjustment of the others until equilibrium is re-established at a different position within the environment.

(Donaldson, 1985, p.29)

This paradigm is compatible with the flexibility-stability idea described in the previous section in the sense that organisations are required to sustain stability while being flexible in order to adapt to external changes. However, the idea of a flexibility-stability dilemma requires a more dynamic version of the structural-functionalist approach which does not imply that we are dealing with social systems that find themselves in a stationary state.

First, the notion of performance gaps as described above and in chapters 2-3 suggests that changes may be initiated within and not only without the organisation. Performance gaps occur as objective data in the form of deteriorating levels of profit, faulthy production volumes or deficiencies of information processing - but they do also occur in the mind of the organisation members as ideas about how things may be done differently. In turn, these ideas may lead to innovation proposals. The subjective perception of performance gaps tends to be stimulated by the way in which organisation members search for and use information, to the extent that the organisation members value information that it not directly relevant to decision making.[9]

Second, although changes may appear to be exogenous from the point of view of the individual organisation, changes do not come about in an automatic fashion as manna from heaven. Changes in the competitive environment are initiated and instituted by the organisations within that environment, and in many cases organisations even organise themselves in order to innovate, i.e. infuse changes to the environment. This idea is prominent in the large number of studies where it is possible to find a classification of organisations according to their ability to innovate and exploit their competitive strengths

---

9. Feldman & March (1981) argue that this is often the case: "First, organizations provide incentives for gathering extra information. These incentives are buried in conventional rules for organizing (e.g. the division of labor between information gathering and information using) and for evaluating decisions. Second, much of the information in organizations is gathered and treated in a surveillance mode rather than a decision mode. Organizations scan the environment for surprises as much as they try to clarify uncertainties. Third, much of the information in organizations is subject to strategic misrepresentation. It is collected and used in a context that makes the innocence of information problematic. Fourth, information use symbolizes a commitment to rational choice. Displaying the symbol reaffirms the importance of this social value and signals personal and organizational competence" (ibid., p.182).

(e.g. Freeman, 1982; Miller & Friesen, 1982; Pelz & Munson, 1982; Porter, 1980, 1985; Cooper, 1983; Tushman & Nadler, 1986).

Third, the way in which the coordination takes place may, in itself, create disturbances and institute changes. From a structural-functionalist point of view, the structure of the organisation determines the effectiveness of organisational activities. Since we are dealing with a social system, coordination has to take place as interaction between the organisation members. Whatever the type of interaction, it must rely on the processing and communication of information. This description amounts to the cybernetic conception of organisations as a "set of elements linked almost entirely by the intercommunication of information" (Zaltman et al., 1973, p.127), and as appears in chapters 2-5 the coordination by processing and communication of information is by no means trivial and entails a number of sources for ambiguity (March & Olsen, 1976) in relation to the nature of organisational activities and roles, and thus the relationships between the organisation members. *Cybernetic*

Thus, if organisations are social systems, the equilibrium of organisations implied by a structural-functionalist approach is a dynamic social equilibrium. *Dynamic* means that the equilibrium rests on forces which tend to change, either as a response to external changes or by their own momentum. *Social* means that the organisational activities take place as interactions between the organisation members which by their actions create a social context for one another.

In the field of organisation theory, the social context has been interpreted in various ways. For instance, Simon (1957) argues that if organisation members are to accept their organisational roles, the tasks which they have to perform and the way in which authority relations are structured must fall within the area of acceptance of the organisation member. During the 1950-60s, the idea of an informal organisation, which serves the needs of organisation members not fulfilled by the formal organisation, gained acceptance and was promoted by the development of the human relations school, and in their work on organisation development, French & Bell (1973) proposed the existence of an organisational iceberg which consists of formal (overt) and informal (covert) aspects.[10] Some have proposed that organisations may be percieved as paradigms in the sense that organisations are social constructions of reality (e.g. Brown, 1978), and Meyer & Rowan (1977) have even interpreted institutional rules as myths by which organisations create

organisational theory
community of practice

----

10. The formal, overt aspect comprise goals, technology, structure, skills and abilities, and financial resources, while the informal, covert aspect comprises individual attitudes, values, feelings (anger, fear, despair etc.), interactions and group norms (French & Bell, 1973, p.18). They have continued working along these lines, and a fourth edition appeared in 1990.

legitimacy and acquire resources. In recent years, the idea of organisational culture has disseminated in the field of organisation studies, e.g. in the work of Zucker (1977), Pettigrew (1979) and Schein (1985).

The present study interprets the social context in terms of the organisational repertoire defined as the set of organisational routines (Nelson & Winter, 1982). Organisational routines may be described in terms of individual skills, but the individual skills become "meaningful and effective only in some context, and for knowledge exercised in an organizational role that context is an organizational context" (ibid., p.105). Nelson & Winter (1982) characterise skills in the following way: *Routines & Skills*

> In the first place skills are programmatic, in that they involve a sequence of steps with each successive step triggered by and following closely the completion of the preceding one. Second, the knowledge that underlies a skilful performance is 'in large measure tacit knowledge, in the sense that the performer is not fully aware of the details of the performance and finds it difficult or impossible to articulate a full account of those details. Third, the excercise of a skill often involves the making of numerous "choices" - but to a considerable extent the options are selected automatically and without awareness that a choice is being made.   *tacit*
>
> (Nelson & Winter, 1982, p.73)

This conception of a structural-functionalist model of dynamic social equilibrium in organisations is sufficiently broad to capture the range of theories employed in this monograph. While chapter 2 on the flexibility-stability dilemma presents a model which from a sociological point of view argues that the processes which leads to organisational equilibrium are determined by contradictory forces, chapter 3 on the behavioural theory of the firm employs a more simple structural-functionalist model of organisations as adaptive rational systems in the sense of Cyert & March (1963):

> 1. There exist a number of states of the system. At any point in time, the system in some sense "prefers" some of these states to others.
> 2. There exists an external source of disturbance or schock to the system. These schocks cannot be controlled.
> 3. There exist a number of decision variables internal to the system. These variables are manipulated according to some decision rules.
> 4. Each combination of external schocks and decision variables in the system changes the state of the system. Thus, given an existing state, an external schock, and a decision, the next state is determined.
> 5. Any decision rule that leads to a preferred state at one point is more likely to be used in the future than it was in the past; any decision rule that leads to a nonpreferred state at one point is less likely to be used in the future than it was in the past.
>
> (Cyert & March, 1963, p.99)

Chapter 4 on the contingency theory of organisational action extends this perspective and employs a model which combines the perspectives of chapters 2-3 and describes the structure of the organisation in terms of a match between the uncertainty of the external environment and the intraorganisational diversity of tasks, goals and perceptions. Chapter 6 argues that search for solutions to performance gaps may be incomplete and subjected to ambiguity, and that performance gaps may, in some cases, never be closed. As will appear from these chapters, the elaborated version of the structural-functionalist approach described in this section is able to embrace these various perspectives. *Novelty*

Until now, the description of the structural-functionalist approach has stressed that the resolution of the flexibility-stability dilemma is far from trivial. The reason for this is that the requirements of the two stages mentioned previously, the stage of initiation and the stage of implementation, are different. At the initiation stage, the organisation must gather and process new knowledge associated with the generation and evaluation of proposals for innovation. At the implementation stage, the organisation must institute rules of procedures. However, the description of the simple idea above raises an important question: What is meant by innovation? While chapters 3 and 6-7 discuss the concept of innovation in some detail, a more simple and straightforward definition may suffice at the present moment. Innovation is defined as a new way of doing things, which means a change in the coordination of organisational activities, in process technology, or in product technology. Innovation may vary according to the degree of novelty. The degree of novelty is conventionally, in the fields of organisation theory and innovation economics, described in terms of a radical-incremental, or major-minor, continuum. Radical innovation refers to substantial changes, and incremental innovation to less substantial, often gradual changes. While radical innovation represents an idea which is new to the organisation, incremental innovation represents an idea which is more obvious. As opposed to incremental innovation, radical innovation entails a large degree of uncertainty regarding the outcome of the process. This means that the search for and processing of information tends to be more extensive in the case of radical innovation than in the case of incremental innovation. *information + radical*

Initiation, implementation and the transition from initiation to implementation requires that the organisation members learn about new opportunities and the way in which these opportunities can be exploited. Thus, innovation depends on learning. As argued by Lundvall (1992a), learning depends primarily on interaction and is, thus, a socially embedded process. As appears in chapters 3-6, learning may take a number of forms, but in any case the social embeddedness implies that learning may become distorted. The risk of learning becoming distorted is larger in the case of radical innovation than in the case of incremental innovation: While incremental innovation represents a minor deviation

22

— Definition of innovation
— When do teachers learn?

from the existing routines of the organisation, radical innovation represents a break which tends to make the existing routines obsolete. Radical innovation implies to a greater extent than incremental innovation the phenomenon of *exnovation*, i.e. the removal of existing routines from the organisational repertoire. In consequence, the success of radical innovation is more sensitive to the social context than the success of incremental innovation. *obsolete routines*

Learning may be more or less formalised. As appears in chapter 3, innovation may be programmed or non-programmed. During the recent years, it has often been argued that innovation is increasingly becoming dependent on formalised search based on scientific research (Dosi, 1988; Teece, 1988). Furthermore, it has been argued that the innovation process differs across industries and sectors (Pavitt, 1984), and that the nature of learning and search changes as the diffusion and implementation of microelectronics and information technology promote new management principles (Gjerding, 1992, 1992a). Chapters 7-9 deal with the issues of sectoral dependency and changing management principles, but the issue of research dependency is not a topic of the present study.

learning

1.4. The compatibility of the organisation theories

Although it is possible to embrace the organisation theories employed in chapters 2-4 within the framework of the elaborated structural-functionalist approach, it should be recognised that the theories represent *different* perspectives on the study of organisational action.

Following Pfeffer (1982) and Scott (1992), the discussions in chapters 2-4 may be described as an attempt to combine rational and natural system models in an open system perspective. Pfeffer (1982) distinguishes between theories according to their perspective on action and the level of analysis, while Scott (1992) distinguishes between theories according to whether they hold a perspective of organisations as rational or natural, and closed or open systems. Figure 2 outlines the Pfeffer (1982) distinction, while figure 3 compares the rational and natural perspectives.

As argued by Pfeffer (1982), the perspective on actions as purposive, intentional, goal directed and rational implies that actions are undertaken as the consequence of choice directed towards goals. Actions may be more or less maximising, i.e. maximising as in the mainstream economic models, or satisficing (Simon, 1957) as in the behavioural theory of the firm. The perspective on actions as externally constrained and controlled implies that actions are taken, reactively or proactively, in response to pressures, the control and

maximizing
satisficing - theory of the firm

23

sometimes even the recognition of which lies outside the scope of the organisation.[11] The perspective on actions as emergent phenomena implies that action cannot be explained as "either an internally directed or an externally determined rationality of behaviour" (Pfeffer, 1982, p.9), but depends on the gradual evolution of the organisational context in which the organisation members find themselves[12], e.g. through a process of institutionalisation where social relationships become embedded in the way in which organisation members percieve the task environment.   *rational n natural*

According to Scott (1992), the distinction between the perceptions of organisations as rational or natural systems relies on whether one describes organisations as designed artifacts or natural phenomena, cf. figure 3. Both perspectives define organisations as collectivities, but whereas the rational system perspective emphasises the formalisation of organisational behaviour towards the attainment of relatively specific goals, the natural system perspective percieves organisations as organic, or living systems (Morgan, 1986): The organisation is a collective of organisation members who "share a common interest in the survival of the system and who engage in collective activities, informally structured, to secure this end" (Scott, 1992, p.25, italics omitted). The main difference between the rational and the natural system models may be described as a difference in the relative importance attached to goal directed behaviour, i.e. a different attitude to the possibility of deliberately designing organisations as devices for goal attainment. While the rational system perspective is primarily normative and concerned with organisation design issues aimed at organisational effectiveness measured in terms of goals and objectives, the natural system perspective is primarily descriptive and concerned with the power relations and processes of interaction which shape organisational behaviour. In consequence, sociological issues pervade the natural system analysis to a higher degree than the rational system analysis.   *Power in natural systems*

According to the theoretical taxonomies proposed by Pfeffer (1982) and Scott (1992), the combination of organisation theories, which is discussed in the present study, entails the perspectives of organisations as (1) open rational systems at the organisational micro and macro levels, and (2) open natural systems at the organisational macro level, cf. figure 4. This combinations raises at least two methodological issues in relation to the concept

---

11. Pfeffer & Salancik (1978) is an example of this perspective, which is not addressed in the present study.

12. Examples of this perspective in the present study are Cohen, March & Olsen (1976) and March & Olsen (1976, 1976a).

Figure 2. *Perspectives on organisation theory*

Level of analysis	Perspectives on action		
	Purposive, intentional, goal directed, rational	*Externally constrained and controlled*	*Emergent, almost-random, dependent on process and social construction*
Individuals, coalitions, or subunits	Expectancy theory Needs theories and job design Political theories	Operant conditioning Social learning theory Socialisation Role theories Social context effects and groups Retrospective rationality Social information processing	Ethnometodology Cognitive theories of organisations Language in organisations Affect-based processes
Total organisation	Structural contingency theory Market failure/transaction costs Marxist or class perspectives	Population ecology Resource dependence	Organisations as paradigms Decision process and administrative theories Institutionalisation theory

Source: Pfeffer (1982, p.13), table 1.1

Figure 3. *Rational and natural views of organisations*

	Rational model	Natural model
	The organisation as a designed artifact	*The organisation as a natural phenomenon*
Basic question asked	How can organisational effectiveness best be achieved?	How can organisational behaviour best be explained?
Existence of goals	Goals exist, but they may be multiple, conflicting and perhaps displaced	Goals are an inappropriate concept. Behaviour is better explained in terms of power and processes of interaction
Control	Overall guidance of the organisation towards objectives	Exercise of power and influence by groups
Major stress	Formal organisation Organisation design	Informal organisation Unanticipated results
Orientation	Normative	Descriptive
Role of management control	Rational and neutral procedures used to help ensure overall effectiveness	Tools used by one group to enable them to dominate other groups

Source: Adapted from Otley (1988, p.88), table 5.1

of organisations held by the different theories, namely the problems of bridging (a) rational system models at the organisational micro and macro levels, and (b) rational and natural systems models at the organisational macro level. These problems are dealt with in the course of the arguments in chapters 2-4, but some general considerations should be mentioned in this introductory chapter.

(a)   Although the level of analysis differ, the rational system models mentioned in figure 4 share the same concept of organisation, i.e. they view organisations as formal goal directed entities. The main differences between the two types of theories are the extent to which the context of the organisation and the preferences at the organisational micro levels are allowed to influence the organisation design, and in the context of the present study the theories are compatible for at least two reasons:

Figure 4. *Analytical perspectives of open system models combined in the present study*

Perspectives on action	Purposive, intentional, goal directed, rational	Emergent, almost random, dependent on process and social construction
Level of analysis	Rational system model	Natural system model
Individuals, coalitions, subunits	Behavioural theory	
Total organisation	Contingency theory	Neo-behavioural theory

First, the behavioural theory of the firm may be viewed as a type of political theories which explain how organisational goals and behaviour are determined as the outcome of competing subgoals and diversity of organisational behaviour. If contingency theory treats the organisational micro levels as a "black box", as implicitly argued by Pfeffer (1982), the behavioural theory illuminates that box and explains how the organisation design in question is a reflection of underlying political processes.

Second, the words "if contingency theory..." were used quite deliberately, because contingency theory does in fact *not* abstain from the discussion of conflicting goals and diversity of behaviour at the organisational micro levels, although the focus is on the total organisation. The issue of interdepartmental conflicts was present in the work by Woodward (1965) described in section 1.1. Burns & Stalker (1961) do describe organisations as political systems of vested interest, and Lawrence & Lorsch (1967) argue that the mechanisms of differentiation and integration are devices for coping with extraorganisational uncertainty by resolving intraorganisational conflicts. The organisation design mechanisms proposed by Galbraith (1977) are not only mechanisms for coping with information overload, as often described, but also mechanisms for resolving intradepartmental conflicts. Finally, Thompson (1967) employs the concept of a dominant coalition in the sense of Cyert & March (1963) and invoke the issue of intraorganisational power in his analysis. Thus, it is often quite difficult to discern where behavioural theorising stops and contingency theorising begins.[13]

---

13. In consequence, I find it difficult to agree with Pfeffer (1982, p.133) when he argues that it "is clear that theories of organization-level rationality are *inconsistent* with the view of

(b)   Although the concept of organisation differs somewhat between rational and natural system models, they may not be incompatible.

First, rational and natural system models are both found within the perspective of organisations as closed or open systems[14], and Scott (1992, pp.100-22) actually argues that the development of organisation theory during the present century has occurred in the form of a gradual transition of the rational and natural system models from a closed system perspective to an open system perspective.

Second, the rational and natural system models may be integrated within an open system framework, either at the extraorganisational level as in Lawence & Lorsch (1967) who distinguish between different organisations which exhibit the characteristics of either rational or natural systems depending on their environmental context, or at the intraorganisational level as in Thompson (1967) who argues that all organisations are mixed rational and natural systems depending on the intraorganisational level of analysis.

1.5. Overview and summary of chapters 2-9

The ten chapters of this monograph may be grouped into four parts which, hopefully, are integrated: The present chapter 1 outlines the basic approach and presents a number of methodological issues, while chapter 10 discusses the extent to which the present monograph has been successful in the tresspassing of organisation theory and innovation economics; chapters 2-5, which reinterpret the flexibility-stability dilemma in terms of some classic organisation theories, deal mainly with organisational and process innovation and abstract from the distinction between imitative and non-imitative change and sectoral patterns of innovation; chapters 6-7, which take these factors into consideration, represent an attempt to transcend the borders of these classic organisation theories and present the reinterpretation of the flexibility-stability dilemma in terms of processes of learning associated with technical innovation; finally, chapters 8-9 delve into the challenges to organisation design posed by the diffusion and implementation of advanced manufacturing technology associated with microelectronics and information technology.

---

organizations as coalitions" (italics added), although I do agree that the type of analysis and the issues at the focus of attention of the researcher are different in the two cases. However, these differences do not, necessarily, imply inconsistency between theories.

14. The closed and open system perspectives may be differentiated by the degree to which uncertainty is allowed to enter the analysis of the organisation (Thompson, 1967).

Chapter 2. The Flexibility-Stability Dilemma

Chapter 2 on the flexibility-stability dilemma describes the antecedents of that notion (Zaltman, Ducan & Holbek, 1973), which later has been restated as the innovation design dilemma (Holbek, 1988). This description is undertaken in three steps. *accountability system?*

(1) Section 2.1 presents a simple, but suggestive, model of intraorganisational dynamics within a large manufacturing firm. The model is based on the idea that organisational action is triggered by the occurrence of performance gaps, and that the process which leads to the closure of the performance gaps may be described in terms of the change of a set of organisational goals. The set of organisational goals represents a compromise between the goals of the intraorganisational subunits, i.e. a balance of the constituent goals each weighted by the amount of intraorganisational power possessed by the subunits in question. This power balance is dynamic in the sense that it rests on processes of learning which persist even though a balance has been established, and the balance tends to be continuously upset as the outcome of interaction between learning, operating procedures and the formation of organisational goals.

(2) The nature of this interaction is illuminated by section 2.2 which discusses the classic flexibility-stability approach represented by Zaltman et al. (1973) and inspired by the work of Hage & Aiken (1970). It is argued that the interaction is iterative and based on the overlap between the stages of initiation and implementation of change. Initiation refers to the process whereby the organisation members become aware that a performance gap exists and begin searching for solutions which may be effective in closing the performance gap. Implementation refers to the institution of the changes necessary to close the performance gap in a trial-and-error fashion. The recognition of performance gaps, and the way in which initiation and implementation are undertaken, depend on the set of intraorganisational rationalities (cf. figure 9) and their integration, and it is suggested that organisations are more likely to overcome the flexibility-stability dilemma if they are able to differentiate in time or space.

(3) Finally, section 2.3 has some critical remarks on the classic approach. It is pointed out that the flexibility-stability dilemma may have changed during the last twenty years due to a rapid dissemination of advanced manufacturing technology throughout the industrialised economies. The utilisation of advanced manufacturing technology presents the manufacturing firm with some new problems of organisation design which are discussed in chapters 8-9. Furthermore, it is argued that *intra*organisational differentiation solutions may entail a number of conflicts between intraorganisational rationalities, and differentiation in space may be more adequately achieved through *inter*organisational arrangements.

29

satisficing & bounded rationality

Chapter 3. A behavioural theory of organisational action

Chapter 3 describes the organisational repertoire in terms of bounded rationality (Simon, 1957) and interprets the flexibility-stability dilemma from the point of view of the behavioural theory of the firm (March & Simon, 1958; Cyert & March, 1963). After a brief account in section 3.1, which describes the evolution of the behavioural theory as part of a major theoretical opposition to standard neoclassical economics, section 3.2 discusses the concepts of satisficing behaviour and bounded rationality as intertwined concepts. According to the behavioural theory, the problem-solving decision maker applies a simplified model of the world, since he is unable to grasp the vaste amount of information necessary to maximise according to the prescreptions of standard economics, and he chooses among those alternatives which meet or exceed some minimum criteria. This choice is based on problemistic search which is primarily characterised by local rationality. In order to validate this argument, section 3.2 analyses how the behavioural theory determine the organisation members' focus of attention and argues that the main determining factors are the differentiation and persistence of subgoals, the individual's span of attention, and the intraorganisational communication channels and tolerance for interdependence. Figure 11 presents a novel way to depict this argument.

Section 3.3 argues that organisational learning may be defined in terms of the shifting of the focus of attention. These shifts are initiated through problem-oriented search and communication, and reflect the change of organisational routines as a response to performance gaps in the form of a discrepancy between levels of aspiration and levels of achievement. This process is characterised as satisficing rational adaptation dominated by local rationality, which leaves little room for organisational and technical innovation. However, the behavioural theory admits some occasions for innovation, and section 3.3 presents a new ideal-type classification of these occasions in terms of the nature of problem-solving, cf. figure 14. It is argued that innovation related to necessity, stress and opportunity reflects reproductive problem-solving, i.e. search for solutions similar to or in the neighbourhood of existing activities, while innovation related to organisational slack primarily reflects productive problem-solving, i.e. search for solutions which are less similar to existing activities and involves new knowledge.

Finally, the flexibility-stability approach is restated from the behavioural point of view, cf. figure 17 which displays some dissimilarities. These differences are caused mainly by the fact that (1) innovation in a behavioural setting is more of the routine than the non-routine type, and (2) the behavioural theory deals mainly with the stage of initiation.

problemistic search
local rationality?

Chapter 4. Contingent decision making and problem solving Is this the case for education

In terms of the classic contingency theory, chapter 4 elaborates on the flexibility-stability approach developed in the previous chapters. The guiding theme of contingency theory is the attempt to answer the following question: How can integration be facilitated without sacrificing the degree of differentiation which is necessary in order to cope with the extraorganisational task environment? The basic idea is that an effective organisation design reflect the extraorganisational task environment in the sense that the organisation is segmented into subunits designed for coping with certain environmental segments. However, segmentation creates a diversity of intraorganisational rationalities which have to be integrated in order to overcome potential conflicts and achieve coordination. This creates an intricate balance between integration and differentiation which differs across organisations according to the nature of the extraorganisational task environment. Thus, organisations differ according to the contingencies which they try to control, and, in consequence, there is no best way to organise.

Section 4.1 describes this conclusion by reference to the work of Burns & Stalker (1961) and Lawrence & Lorsch (1967). Burns & Stalker (1961) argue that management practices differ across a continuum with organic and mechanistic designs positioned at the opposite ends. Mechanistic designs are mostly found in stable task environments and are, consequently, characterised by hierarchial coordination based on functional specialisation and vertical interaction, while organic designs, mostly found in volatile task environments, are characterised by network coordination based on the continual redefinition of tasks and horisontal interaction. Lawrence & Lorsch (1967) go one step further and argue that the differences in management practices are reflected in the means by which intraorganisatio-nal conflict is resolved, cf. figure 21.   conflict

In terms of organisation design, the seminal work of Burns & Stalker (1961) and Lawrence & Lorsch (1967) indicate that the natural system model assumes predominance over the rational system model as the extraorganisational task environment becomes more volatile. However, this does not imply that the organisational configuration tends to adhere to the types of designs suggested by the natural system approach. Following Thompson (1967), and in accordance with Burns & Stalker (1961) and Lawrence & Lorsch (1967), section 4.2 argue that the most viable organisations exhibit a range of intraorganisational rationalities at the level of departments according to the nature of the departmental task environment. Thompson (1967) suggests that the rational and natural system models reflect, respectively, a closed-system and open-system approach which are not only found in the scientific study of organisational behaviour, but are parallelled by similar approaches in management practices, as well. Thus, organisations try to impose a closed-system logic on the core of their productive activities, i.e. a condition near

organic and mechanstic designs, horisontal + vertical

certainty where as many contingencies as possible have been removed and delegated to other parts of the organisation which, in consequence, tend towards an open-system logic. This tendency is positively correlated with the amount of contingencies that enter the decision domain and implies an increase in the amount of communication and the frequency of decisions which have to be undertaken.

However, although a combination of closed and open system logic intuitively seems effective for organisational performance, the management may find it difficult to reconcile these two types of logic. A closed-system approach implies that management seeks to eliminate uncertainty, while an open-system approach implies that management strives for flexibility by freeing organisational resources to cope with uncertainty. In the short run, a closed-system approach may be effective, but in the long run flexibility is necessary in order to secure organisational survival. Thompson (1967) refers to this flexibility-stability problem as the *paradox of administration* and suggests that it may be handled through an alternative to behavioural problemistic search, i.e. opportunistic surveillance which is a proactive search mode that tries to anticipate new solutions and does not "stop when a problem solution has been found" (ibid., p.151). Section 4.2 suggests that the paradox of administration may be analysed in terms of a conflict between computational and non-computational strategies, cf. figure 24, and makes the novel proposition that opportunistic surveillance may imply reproductive as well as productive search.

This way of solving the flexibility-stability dilemma implies that uncertainty, from a contingency point of view, appears as an information gap, i.e. a discrepancy between the knowledge possessed by the organisation members as a collective and the knowledge necessary to cope with contingencies. Following Galbraith (1977), section 4.3 describes a number of organisation design strategies for increasing the information-processing capability of the organisation and, especially, discusses the design for creating lateral relations. It is argued that the success of lateral relations in terms of organisational performance depends on the coordination of a diversity of goals and perceptions, i.e. rationalities, which are brought together in a task environment. Since the creation of lateral relations implies an increase in the number of attention centres which have to interact, it also implies an increase in the amount of potential organisational conflicts. In consequence, the contingency solution to the flexibility-stability dilemma is based on an intricate balance at the knife-edge between forces that may be either dysfunctional or synergetic.   *information processing capability*

Chapter 5. Two cases of organisational innovation
Applying the restatements of the flexibility-stability dilemma provided by the previous chapters, chapter 5 describes two cases of organisational innovation which have been

opportunistic search

analysed and carried out in an action research fashion by a group of consultants to which I belonged. Inspired by the Morgan & Smircich (1980) taxonomy on ontological assumptions in social science, section 5.1 preliminary suggests that the behavioural and contingency theories belong to the same body of theory in the sense that they both reflect rational system models and focus on organisational processes of information processing and decision making. A combined behavioural-contingency approach characterises the way in which the two cases of organisational innovation reported in chapter 5 were undertaken by the consultants who assisted the two manufacturing firms involved in the project. In both cases, organisational innovation was initiated by the management in order to institute a new way of strategic planning which would lead to product innovation with less lead time and more efficient use of existing resources. In firm A, the changes implied an entirely new way of planning based on teamwork and reorganisation of the task environment, and consequently firm A may be characterised as a case of radical innovation. In firm B, the changes were less pervasive and implied some minor changes of planning routines based on interdepartmental communication, and hence we are dealing with a case of incremental innovation.

In both cases, a number of problems occurred which tended to stall the process of innovation at the initiation stage. As described in section 5.2, the organisational innovation was based on an increased frequency of communication and feedback between attention centres which were forced to engage in, and continously change, a common task environment, and this process created a number of organisational and human barriers to change. As could be expected, these problems were more difficult to overcome in firm A, i.e. the case of radical innovation, and the way in which these problems were solved implied an action research strategy based on psychological guidance of the organisation members. While the organisation members in firm B were introduced to various methods of planning and communication through teaching and discussions, the organisation members in firm A were forced to find their own solutions within the framework of a number of newly established teams and recieved therapeutic assistance during this process. Because the radical innovation disturbed the balance between the organisation member's professional and social roles within the organisation, the therapeutic assistance focussed primarily on how the organisation members could express their personal problems and insecurities, and aid one another to overcome these problems, partly by developing team-oriented skills.

Section 5.3 describes the actions undertaken and argues that the succes of the organisational innovations in both cases was positively correlated with the extent to which the organisational and human barriers to change were made visible to the participants. The innovation was successfully carried out because the consultants managed to provide three

making barriers to ∆ visible

33

types of visibility: (1) Personal visibility, which refers to the ability of the organisation members to display their feelings of insufficiency and provide assistance for one another through interaction; (2) strategic visibility, which refers to the responsibility of managers to articulate and communicate the strategic vision of the organisational innovation, and the responsibility of the subordinates to engage in that process; and (3) competence-oriented visibility, i.e. the ability of the organisation members to apply their social and professional skills in an entirely new task environment.

Finally, section 5.4 discusses the evidence on the flexibility-stability dilemma provided by the two cases of organisational innovation and makes the following six suggestions: The organisational innovation in firm A stalled at the stage of initial implementation, which is the transitory stage between initiation and routinisation of the new organisational activities; both cases reflect the problem of uncertainty in the Galbraithian sense of an information gap; the changes undertaken in firm A resulted in the creation of more complex attention centres which enhance the ability of the organisation to overcome the flexibility-stability dilemma; in both cases the tolerance for interdependence was increased through the creation of self-contained tasks; however, the organisational innovation in firm A may invoke contradictory forces because the creation of more complex attention centres implies an increasing diversity of intraorganisational rationalities at both the intra- and interdepartmental levels; finally, while the changes in firm B were undertaken in order to facilitate reproductive search, the changes in firm A use reproductive search in order to develop an agenda for opportunistic surveillance.

Chapter 6. Learning cycles and some concepts

Chapter 6 elaborates on the perspectives of chapters 3-4 by discussing barriers to organisational learning and outlines some concepts on learning processes. Organisation theory and innovation economics have developed different learning concepts, and the concluding section 6.5 discusses some intersections and proposes a novel taxonomy which may be used for future research purposes.

Following Hedberg (1981), section 6.1 argues that it is meaningful to use the concept of organisational learning, although only organisation members, not organisations, are able to learn. The learning of organisation members takes place within the social context of the organisation, and the outcome is reflected in various means of storing knowledge, routines and behavioural regularities which are something more than the cumulative result of individual learning. Hedberg (1981, p.9) suggests that the outcome of organisational learning tends to shift "inside behavioural modes", which he describes as (1) adjustment learning that refers to the change of parameters and rules, (2) turnover learning, which refers to the change of the set of organisational routines by adding new ones and removing

34

some of the old ones, and (3) turnaround learning which denotes the changes of the theories of action (Argyris & Schön, 1978) held by the organisation members. However, as described in section 6.2 on the neo-behavioural approach (Cohen, March & Olsen, 1976; March & Olsen, 1976, 1976a), organisational learning may be impeded for a number of reasons: *First*, a direct link between individual beliefs and action may be absent. This is the case of role-constrained learning; where the individual action is confined within a set of organisational routines that makes a translation of individual beliefs into individual action impossible. *Second*, even if individual action actually is changed as the outcome of individual learning, there may be little or none impact on organisational action. This is the case of audience learning, where the individual is unable to influence the way in which the organisation attend to and solve problems. *Third*, even if organisational action is changed, it may reflect a perverse relationship between organisational action and the extraorganisational task environment. This is the case of superstitious learning, where the organisation members fail to interpret environmental responses to organisational action, or interpret them erroneously. *Fourth*, and finally, the organisation members may be unable to interpret or form an unanimous opinion on subsequent environmental responses. This is the case of learning under ambiguity where "there exists no single, objective explanation of an outcome or of its causes" (Hedberg, 1981, p.11).

The bounded rationality approach has found its way into various types of analysis. Section 6.2 presents a number of arguments in favour of the proposition that the neo-behavioural learning concepts may be regarded as a broader version of the concepts of bounded rationality and satisficing behaviour; hence the label *neo-behavioural*. Section 6.3 applies the bounded rationality approach to evolutionary theory in the sense of Nelson & Winter (1982), who are inspired by the behavioural theory of the firm. It is argued that the ability of the organisation to deal with performance gaps in the Nelson & Winter (1982) setting depends on the nature of the internal and external selection environments, and the interaction between these environments, the dynamics of which was previously described in the preface. The surviving routines tend to persist, for at least three reasons: The propensity to employ precisely those routines which proved valuable in the past is high; the agenda for search is dominated by standard operating procedures; and the intra- and extraorganisational task environments tend towards mutual adjustment. However, contradictory forces are at work, since learning processes tend to upset the tendency towards mutual adjustment by the occurrence of innovations in the extraorganisational environment and by the changes of the organisation members' perceptions. Furthermore, standard operating procedures tend to change due to learning by doing in the Arrow (1962) sense, i.e. the accumulation of experience in localised problem solving when the

behavioral theory of the firm + evolutionary theory   *Arrow* *

operating procedures have become standard. Following this approach, section 6.3 employs some Danish data on manufacturing productivity and R&D during 1974-90 in order to test the hypothesis that operating procedures provide the agenda for the search for new solutions. It is suggested that if learning in research and development is localised, the transition from search to R&D takes place at a high speed, and *vice versa* in the case of less localised search. The test provides some support for this hypothesis. Finally, it is argued that while learning by doing results in localised search, search becomes less localised in the case of learning by using (Rosenberg, 1982), i.e. learning through feedback from innovation users. *learning*

The set of organisational routines defines the path along which organisational learning takes place and provides the organisation members with a socially ordered context in which individual learning takes place. This implies that learning is an interactive process subjected to behavioural regularities. Applying the model of Johnson (1992), cf. figure 32, section 6.4 discusses these regularities from an institutional point of view, and describes a number of learning concepts which may be ranked according to an increasing frequency of human interaction. At the high-frequency end, one may find learning by searching which refers to types of profit-oriented learning, and learning by exploring, i.e. search activities in primarily non-profit organisations. In the middle range one may find learning by producing which denotes processes of learning associated with the normal production activities of the organisation. Learning by producing comprises learning by doing, learning by using, and learning by interaction which refers to the outcome of feedback between producers and users. Although learning by interacting was, originally, proposed by Lundvall (1985) as an interorganisational concept, section 6.4 argues that the concept may be used at the intraorganisational level as well.

These concepts have been developed in the field of innovation economics. In the field of organisation theory, Argyris & Schön (1978) have proposed the concepts of single-loop and double-loop learning, which may be used at a more general level in order to characterise the types of learning described so far. While single-loop learning implies the change of routines within a constant framework of norms and performance, double-loop learning refers to the case in which this framework has to be changed in order to overcome performance gaps. Section 6.4 argues that to the extent that organisations learn to master both single- and double-loop learning, we may talk about something broader than learning to learn (Stiglitz, 1987), i.e. deutero-learning (Bateson, 1942).

Finally, section 6.5 proposes a novel taxonomy whereby learning by producing, searching and exploring may be analysed in terms of the neo-behavioural concepts of audience, role-constrained and superstitious learning, and learning under ambiguity. It is argued that the behavioural concepts may be applied to all of the concepts developed in

institutional point of view
synergy w/ innovation economics

the field of innovation economics, except from the case of learning by doing in which only role-constrained learning can occur.

Chapter 7. Paradigms and activities in the innovation process
Chapter 7 discusses innovation as an interactive process and presents a number of models which are useful in order to determinate the direction of search and focus of attention in the innovation process. These models have been developed in the field of innovation economics, but are interpreted within the flexibility-stability approach of the previous chapters. The level of analysis in section 7.1 is the innovative manufacturing firm. Section 7.2 elaborates on this perspective and, furthermore, discusses interrelations within the production system. Finally, section 7.3 presents some generalised stylised facts on innovation and discusses long-term changes at the level of the economy.

Preliminary, section 7.1 discusses a classification of models presented by Saren (1984), cf. figure 35, and argues that although each of these models are able to yield important insights, they all suffer from the fact that they apply a sequential nature of analytical reasoning and regard the innovation process as a linear progression which only to a limited extent captures the element of learning through feedback. Instead, the chain-linked model proposed by Kline & Rosenberg (1986) is advocated as a possible starting point for a more general model. The chain-linked model, as depicted in figure 36, comprises a central core of progressive stages interlinked by feedback from testing, design, production and marketing which occur as a dominant design evolves. Section 7.1 suggests that the chain-linked model may be elaborated in terms of the Saren (1984) classification and applied to specific analytical cases by identifying the formal task environments, and the occurrence of stimuli, activities and decision points that occur during the innovation process. Such an analytical combination would have to include the existence of routines, routines for changing routines and intraorganisational conflicts on routines in order to capture the insights provided by the previous chapters. In consequence, section 7.2 presents two models described by Christensen, J.F. (1992) and compares them to the classic flexibility-stability approach presented in chapter 2, cf. figure 38. These models comprise the intraorganisational conflicts on command of internal and/or external resources for innovation and discusses from a sociological point of view the types of organisational changes which accommodate product innovation. Thus, they provide a useful complementary to the approach presented in the previous chapters which have mainly dealt with organisational and process innovation.

Section 7.2 describes some patterns of innovation at the levels of the manufacturing innovating firm and the production system. The main purpose of this section is to elaborate on the strategy approach suggested by Freeman (1982). Freeman (1982) argues

chain linked model of innovation

Freeman

that in order to provide an interpretation and understanding of innovative behaviour alternative to mainstream economics, the analyst may delve into "the various *strategies* open to a firm when confronted with technical change" (ibid., p.169). Consequently, he describes six different strategies, cf. figure 40, which "should be considered as a spectrum of possibilities, not as clearly definable pure forms" (ibid., p.170). Section 7.2 describes this spectrum as a *first approach* and suggests that a *second approach* might be obtained by supplementing the strategy approach with an analysis of the match between the intra- and extraorganisational task environment. It is argued that the cultural evolutionary perspective of Burgelman & Rosenbloom (1989) provides such a second approach. Burgelman & Rosenbloom (1989) describe strategy making as a social learning process which balances mechanisms that generate technical capabilities and mechanisms that integrate these capabilities, and they suggest that "successful firms operate within some sort of harmonious equilibrium of these forces" (ibid., p.20). The equilibrium is brought about through the interplay between the internal and the external selection environments. Finally, section 7.2 suggests that a *third approach* for future research may combine the work of Cooper (1983) and Pavitt (1984). In order to characterise 58 innovation projects undertaken by 30 industrial firms, Cooper (1983) distinguishes between three stages of innovation: Marketing, evaluation of the project, and the technical/production stage which comprises design, testing of prototype and production. By measuring the amount of project time spent at each stage, Cooper (1983) arrives at seven clusters, cf. table 3, and like the Freeman (1982) spectrum, the Cooper (1983) clusters may be described as a way of describing the overall innovation process in terms of the *focus* of the innovation strategy. While Cooper (1983) was unable to find sectoral patterns of innovation, Pavitt (1984) suggests that such patterns actually exists. Reviewing about 2,000 innovations in terms of technological opportunities, demand conditions and means of appropriability available to the firms, Pavitt (1984) distinguishes between four types of sectoral clusters, cf. figure 41: Supplier-dominated firms, science-based firms, scale-intensive firms, and specialised equipment suppliers. While Cooper (1983) identifies the main stages of activities in the innovation process in a way that determines how the various activities overlap according to the capabilities which characterise each cluster, Pavitt (1984) identifies these capabilities and explains how they are linked with technological opportunities, demand conditions and means of appropriability. Thus, a combination of the Cooper (1983) and Pavitt (1984) approaches would complement the Freeman (1982) strategy approach and make it possible to interpret the strategy approach from the perspective of Burgelman & Rosenbloom (1989).

Section 7.3 turns to the issue of generalising stylised facts on innovation along the lines suggested by Nelson (1992). Nelson (1992) argues that the analysis of technical progress must be based on an understanding of the technical capabilities of firms from the point of view that firms and environments are co-evolving. He implies that such a dynamic approach might be helpful in translating the conclusions from empirical findings into a formal and more general theory on technical innovation. At the level of stylised facts, several attempts made during the last decade qualifies according to the approach of Nelson (1992), and section 7.3 reviews the contributions of Dosi (1988), Arthur (1988) and Freeman & Perez (1988) which are comparable in the sense that they imply that technical innovation travels along a path that may be recognised as a paradigm in the Kuhnian sense. This path is cumulative in the sense that it sets the direction for future search because it represents an accumulation of endowed resources and knowledge, which exerts an exclusion effect on solutions and implies complementaries between the economic agents at the levels of the firm, the sector, the production system, and the economy. Freeman & Perez (1988) suggest that this effect applies to the entire social fabric and propose the notion of *techno-economic paradigms* in order to describe the match between the economic and social systems at the level of the economy. They identify long waves of technological upsurge each characterised by some key factor of technical and social progress, and they argue that the industrialised economies, at present, find themselves at the beginning of an upsurge based on microelectronics and information technology. Concludingly, section 7.3 suggests that this new upsurge may be characterised by information-intensive production activities which provide economies of scope and economises on all types of productive inputs.

Chapter 8. Implementation of high technology

Chapter 8 discusses the organisational and technical changes associated with microelec-tronics and information technology embodied in advanced manufacturing technology, and argues that a systemic approach is favourable to the competitiveness of firms which employ advanced manufacturing technology. The focus is on process innovation, and the main implication is that the competitiveness of the firm is increased through a *match* between technical and organisational capabilities. Thus, the most competitive firms are not, necessarily, those which make the most capital intensive investments in advanced manufacturing technology.

Section 8.1, which defines advanced manufacturing technology (AMT) in terms of computer aided manufacturing and computer aided design and engineering, outlines the competitive opportunities and organisational problems entailed in the implementation of AMT. It is argued that the new technological paradigm focusses on programmability and reprogrammability of manufacturing techniques instead of hardware control by automated production capabilities. In consequence, economies of scope become an option to firms which employ AMT because they have the opportunity to overcome the trade-off between efficiency and flexibility which characterises types of technology where human skills are built into hardware control. Based on research on computer integrated manufacturing and flexible manufacturing systems, section 8.1 suggests that economies of scope is associated with the realisation of a number of flexibility options, as depicted in figure 43. It is argued that the realisation of economies of scope implies that overall production planning tends to become more centralised, while production operating becomes more decentralised. Furthermore, product development tends to become more customised, and, in consequence, learning by using (Rosenberg, 1982) and learning by interacting (Lundvall, 1985) increases in importance as a source of competitiveness. However, the realisation of the flexibility options may face a number of problems which relate to increasing capital intensity, the necessity of providing a computerised network of communication, an increasing importance of human ressource management, and the simplification of organisational hierarchies (Gupta, 1988; Bessant & Buckingham, 1989).

Section 8.2 discusses the issues of human resource management and organic principles of organisation by comparing empirical evidence from the British, American, French, Italien, German and Swedish industry, and argues that the realisation of the flexibility options takes place as organisational learning through creative tensions. The notion "creative tension" has been applied by Leonard-Barton (1988) in order to describe the occurrence of a disequilibrium between technology and structure which provides the stimuli for change. She argues that organisations are able to learn faster about technical issues than organisational issues, cf. figure 48, because knowledge about organisational issues is, to a larger degree, tacit, and, to a smaller degree, facilitated by organisational arrangements. Section 8.2 provides some evidence on the fact that organisational changes are more difficult to initiate and institute than technical change. In order to overcome this problem, section 8.2 argues extensively on a comparative empirical basis that the realisation of the flexibility options is sensitive to the introduction of organic principles of coordination based on the integration of technical and human capabilities through human resource management and the achievement of an agreement between management and labour on technical change.

Section 8.3 provides some evidence on the Danish industry which supports the findings and conclusions reported in the previous section. It appears that the flexibility options provide important stimuli for implementing advanced manufacturing technology. Furthermore, the organisational performance associated with the use of advanced manufacturing technology is, in particuliar, sensitive to the integration of technical and human capabilities. Concludingly, section 8.3 suggests that larger firms may be in a better position than smaller firms to implement AMT. The availability of financial resources for

40

purchasing AMT provides only a partial explanation. A large production volume is conducive to AMT implementation because it allows the firm to experiment in a piecemal fashion without postponing orders or off-setting production, and a large labour force increases the probability of the firm to possess technical and organisational skills which can be combined in trial-and-error processes.

Finally, section 8.4 takes a look ahead and argues that a larger variety of organisational principles than advocated in sections 8.1-8.3 may occur as advanced manufacturing technology becomes more standardised. According to Freeman (1992b), the industrialised economies are experiencing a strong growth in technical, scientific, managerial and administrative occupations, and section 8.4 argues that the proliferation of these skills is likely to stimulate more organic forms of coordination. However, Freeman (1992b) suggests that this trend may change as information and communication technology become more standardised, and Sundqvist et al. (1988) argue that the incentives to develop a high technical level in the workforce may deteriorate because microelectronics and information technology tends to reduce capital requirements. In the case of CNC machinery, Banke & Clematide (1989) provides empirical findings which indicates a variety of work organisation principles, and Cavestro (1986) argues in a French context that numerical control may contribute to a larger vertical division of labour. Applying the flexibility-stability approach to these empirical findings, section 8.4 hypothesise that firms which apply AMT may encounter creative tensions related to an increasing diversity of intraorganisational perceptions and a proliferation of opportunistic surveillance, and that stress and slack innovation may be promoted

Fordst + Japanese

Chapter 9. New management principles and quality control

Chapter 9 delves further into the future trend of organisational diversity and creative tensions, in two steps: Following Gjerding (1992), section 9.1 compares the ideal-type Japanese and Fordist management systems in order to outline the nature of the flexibility-stability dilemma entailed in the emerging techno-economic paradigm, while section 9.2 argues that total quality management from the Deming (1982) point of view is favourable to the solution of the flexibility-stability dilemma in the new techno-economic setting.

Section 9.1 assumes, as its point of departure, that the dominant management principles of the new techno-economic paradigm may be inferred from the ideal-type Japanese management system, while the previous techno-economic paradigm may be characterised in terms of the ideal-type mass production Fordist management system. This is, of course, a heroic simplification. However, these management systems are invoked in order to analyse, by contrast, some ways of organising the utilisation of advanced manufacturing technology. The approach is ideal-typical in the Weberian sense, i.e. the management

organizational management mismatch

41

systems are described in their pure forms. The description concentrates on the overall management approach, logistic principles, the human resource perspective, the degree of intrafirm specialisation, and the decision making process. Analysing these five aspects, section 9.1 points to the following characteristics of the new management system as compared to the old one: (1) The amount of productive problemistic search and opportunistic surveillance is likely to increase; (2) the types of flexibility-stability dilemmas, which may occur in the new management system, relate to an increasing number of attention centres, an increased differentiation of subgoals, and increasing reliance on coordination through communication; (3) there may exist more opportunities to prevent audience and role-constrained learning, and to stimulate the amount and frequency of feedback. In conclusion, the new management system has to attend to especially four problems: A tendency towards an extensive burden on the incentive system; an insufficient degree of routinisation; information overload; and the rate of implementation of changes.

Section 9.2 takes, as its point of departure, the point of view that the principles of quality management rest on the understanding of manufacturing as a people-driven system. This understanding implies the notion of total quality management (TQM) as reflected in the Deming philosophy, or Deming's fourteen points for management, cf. figure 52. It is argued that the Deming philosophy provides a powerful guide to the establishment of lateral processes that may overcome the new dilemmas described in the previous section, and it is, furthermore, hypothesised that there may exist a positive relationship between the AMT implementation problems described in chapter 8 and a low speed of application of TQM principles. This conclusion and hypothesis are based on a comparison between the use of quality circles in Denmark and Japan (Skyum & Dahlgaard, 1988). The comparison is undertaken in terms of the resources devoted to training for quality management, the focus of attention of quality management, and the involvement of employees, and it is implied that the success of quality management in providing solutions to the flexibility-stability dilemma is sensitive to the organisational capability of the firm to create intraorganisational mutual theories of action.

Chapter 2
The Flexibility-Stability Dilemma

2.1. A simple model of intraorganisational dynamics

Imagine a firm that manufactures and markets some artifacts for industrial purposes, say three variants of a CNC machine. The firm in question is large, in a Danish setting, with 120 employees, and 4 top managers in charge of production & service, sales & marketing, accounting & finance, and research & development, besides a chief executive. The product technology is standardised, however with the possibilities of minor adjustments to specific costumer requirements, and the R&D strategy of the firm emphasises incremental product development. For a period of time, the firm experienced a satisfactory rate of growth in turnover, but during the last four years the annual growth rates have gradually diminished, and the rate of profit have fallen below some minimum requirement defined in the long term business plan that the group of top managers issued six years ago. The rate of growth in turnover and the rate of profit are depicted in figure 5, which shows that the growth of turnover began to decline at the point of time where the long term business plan was issued. In periods 4-6, the decline of the rate of profit was smaller than the decline of the growth in turnover, e.g. due to some kind of rationalisation, lay-offs and minor changes in operating procedures, but during the last two years the decline of the rate of profit has taken place at the same pace as the decline in the growth of turnover. This development reflects that the opportunities within the existing organisational repertoire for off-setting the direct relationship between the growth of turnover and the rate of profit have

Figure 5. *Rates of growth in turnover, rate of profit, and minimum rate of profit required by the firm in question*

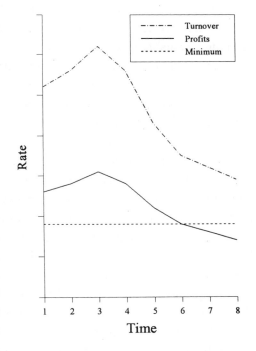

been exhausted. This experience, combined with the fact that the long term business plan and the recession of the economic performance of the firm occurred simultaneously, have created frustration and subsequent intraorganisational tensions at practically all levels of the firm, and these tensions have been growing since the rate of profit fell below the minimum level prescribed by the business plan.

Now, initially the business plan represented a compromise between goals and perceptions at the levels of departments and industrial relations. Before becoming official, the business plan had been discussed throughout the organisation, and conflicting interests between departmental managers, between employees grouped according to trade and departmental affiliation, and between management and employees had been traded off according to the relative strength of intraorganisational power. In consequence, a type of goals and perceptions "equilibrium" had been attained on the basis of a number of minimum requirements associated with each type of goal that could gain some legitimacy *vis-á-vis* the business plan. However, the declining economic performance of the firm, and the various rounds of rationalisation and lay-off, have increasingly threathened the stability of the organisational "equilibrium", and at present in period 8 something *has* to be done. Therefore, the business plan is being revised, and a number of corrective actions are being planned with the purpose of altering the organisational repertoire and institute process and product innovation. Furthermore, the minimum criteria of the rate of profit (and other goals) are reconsidered in the cruel light of experience.

Leaving aside the type of corrective actions undertaken to future discussion in this chapter, we undertake a major travel in time, say to a situation five periods from now. During our travel, we take the opportunity to review the degree to which corrective actions were undertaken in the previous six periods from now and in the future five periods, and we do, somehow, assess the level of deviations from organisational goals measured by an indicator that reflects the relative weight of the performance criteria involved in the business plan. Figure 6, where the deviations from organisational goals are termed *performance gaps*, depicts our review, and the movements of the curves may be described as the "equilibrating" movements of the organisation in question.

Figure 6 is based on a quite simple idea. At some point in time, there exist an affinity between the organisational repertoire and the set of organisational goals as reflected in the business plan, and performance gaps are more or less absent. The exercise of the repertoire tends, in the absence of important performance gaps, towards an increasing degree of dormancy, e.g. due to the fragmentation of organisational routines created by horisontal and vertical segmentation. From the point of view that organisational behaviour is satisficing, the tendency towards dormancy will continue until the set of minimum requirements for organisational goals is violated. The corrective actions undertaken by the firm revive and develop the organisational repertoire and remedy the deficiency of organisational performance. The reason why C and not A or B is the new "equilibrium" position is that at C the level of performance gaps has decreased to the minimum level in question. At A the level of performance gaps is still increasing above the minimum level; at B it is decreasing, but is still above the minimum level. The relative time paths of the two curves illustrate that the level of performance gaps increases as the level of corrective actions decreases, and that corrective actions are undertaken in response to performance gaps. Thus, we are

Figure 6. *Performance gaps and organisational change as movements towards a dynamic social "equilibrium"*

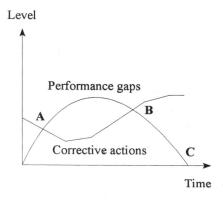

faced with a type of organisational behaviour that tends towards complacence and is reactive in response to the occurrence of performance deficiency.[15] As the level of performance gaps decreases below the minimum level implied by the satisficing behaviour of the organisation, the level of corrective actions tends, once more, towards complacence, and the stage is set for a repetition of the process.

However, any repetition of the process is not static in the sense of interrelated movements that repeat themselves in a mechanical device. Instead, they are dynamic in the sense that the repetition takes place in an organisational setting qualitatively different from the previous setting. The affinity between the set of goals/perceptions and the repertoire involves a mechanism of guiding, where operating procedures and activity programmes define how goals are achieved, while the set of goals define which operating procedures and activity programmes are enacted. During the course of corrective actions, new elements are added to the organisational repertoire and some existing elements may be removed, and the development of the organisational repertoire impacts and reflects a simultaneous alteration of the relative weight between organisational goals. Furthermore, likewise the organisational repertoire, elements are added to and removed from the set of organisational goals and perceptions. These processes within the activity aspect and the

---

15. However, the type of reasoning presented in relation to figure 6 and 7 may well be adapted to proactive behaviour, cf. section 2.3 and chapter 3.

goal-setting aspect of organisational behaviour imply that learning takes place in an atmosphere of intraorganisational struggle.

For instance, take the example of some interdepartmental grievances that often occur in real-life firms.[16] The sales & marketing department argues that the declining economic performance of the firm reflects the neglect to enter a new market, and the department suggests an increase in marketing efforts and R&D. Furthermore, the sales & marketing department points to failures of delivery due to the fact that the production department neglects the time schedules implied by incoming orders. The production department, on the other hand, explains the declining economic performance by failures of delivery creating disappointment at the costumer side, because the firm faces an excess of incoming orders, the deliveries of which have been promised by the sales & marketing department without due consideration to the production capacity of the firm. Furthermore, the production department points to an increasing rate of recycling of product development, since the improvements suggested by the R&D department are not manufacturable at the initial stage of the improvement. The R&D department enters the intraorganisational debate by arguing that the problems related to the manufacturability of product improvements are, partly, caused by resistance within the production department where some technicians tend to oppose any type of change proposal coming from elsewhere in the organisation. Finally, employee representatives point to the fact that rationalisation and lay-offs despite of a promising business plan have reduced employee motivation and diminished the pool of resources devoted to the development of human capital.

Now history may repeat itself in the form of an evaluation and revision of the business plan. By means of coupling intra- and interdepartmental meetings, market prospects are reconsidered on the basis of revised sales forecasts; product calculations are critically evaluated and a plan for augmenting the production capacity is scheduled; some procedures related to the coordination of sales activity, procurement and production planning are changed; a new schedule for the cooperation between the R&D and production departments is devised; long term perspectives of employment security and education and training are outlined by the top management; and finally the accounting & finance department recalculates the budgets and financial requirements. The outcome of this process appears in the form of a new business plan, which outlines the minimum criteria for growth in market shares, turnover and the rate of profit, principles for the relative weight given to lay-offs *vis-á-vis* training and education as a buffer against low levels of activity, and the role that the firm aims to play in relation to the technical

---

16. The following example are inspired by a number of sources such as Woodward (1965), and in a Danish context Gjerding (1990), Jensen & Christensen (1993) and Melander (1994).

46

development in the field of CNC machinery. Furthermore, the business plan describes the main principles for the coordination of sales and production, production and procurement, production and R&D, and the use of forecasting and market monitoring in medium term tactical planning.

During this process, learning has taken place in terms of the adjustment of the perceptions held by the organisation members at the level of units, departments and management. The set of goals has been revised, and resources in terms of man hours, machine hours and financial resources have been allocated. The organisational repertoire concerning the interdepartmental coordination of departmental activities has been changed by augmenting some operating procedures while removing others. In consequence, a new set of relative weights has been attached to the goals and activities of the various intraorganisational groups and units. The departments and organisational members involved in the process have acted as mediators for certain goals and perceptions, and the set of goals and perceptions that has survived this process represents a new balance between intraorganisational interests in affinity with the organisational repertoire.

The process briefly described above may be conceptualised by the image of the organisation as an evolving system, the forces of which are characterised by (1) the ability of the organisational configuration to perform efficiently in order to solve organisational tasks and (2) the ability to change effectively in order to close occurring performance gaps. While the former implies some sort of stability, or inertia, the latter implies flexibility. Thus, in the context of the present hypothetical case story, we are dealing with a dynamic relationship between stability and flexibility. This relationship may present the organisation with a dilemma to the extent that the existing organisational repertoire has become rigid, or dormant, and represents narrow limits to the directions and scope of the perceptions of the organisation members. The dilemma can only be resolved, if the organisation is able to change its configuration to a degree where the performance gaps in question are either closed or redefined. *resolving dilemma*

Figure 7 represents an attempt to model the type of process described above by depicting three variables underlying the movements towards "equilibrium" shown in figure 6. The three variables are (1) learning, interpreted as the accumulation of knowledge, (2) the change of the intraorganisational balance between conflicting goals and perceptions (power balance), and (3) the change of operating procedures as an indicator of the organisational repertoire. Learning may be interpreted partly as a result of the changes in the power balance and the operating procedures and partly as a prerequisite for these changes. Figure 7 suggests that the corrective actions undertaken rely on learning that undermines the existing power balance through changes in the set of goals and perceptions, and that these changes result in changes of operating procedures. The

Variables

power balance change is more abrupt than the changes of operating procedures, since the alterations of the set of goals and perceptions are, normally, subjected to a type of decision making process that is less continuous and more discrete than the type of decision making process involved in the change of operating procedures. Furthermore, figure 6 suggests that the power balance change is, initially, stimulated by changes in operating procedures that alter the perceptions of the organisational agents,

Figur 7. *Organisational learning, and the rate of change in operating procedures and the power balance of intraorganisational interests (goals and perceptions)*

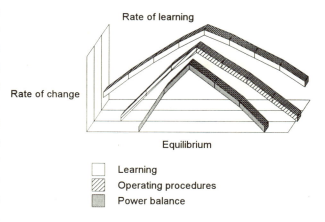

but that this relationship is reversed during the later stages of change in operating procedures, where the power balance assumes the role of guiding principles for the way in which new types of operating procedures are deviced and implemented. The reason why the changes in the power balance and operating procedures are feasible and become effective is that learning takes place at any point in time and continues to do so after the power balance and the new operating procedures are implemented.[17]

power → learny → power

2.2. The classic flexibility-stability approach

The nature of organisational behaviour suggested by the simple model described above is a type of satisficing behaviour, where the recognition of performance gaps stimulates a process of search for alternative solutions, the attachment of possible consequences to the solutions found, and the subsequent decision to embark or not on the solution in question judged by some minimum performance requirement. The model was based on the *interaction* between learning, the formation of organisational goals and the development of operating procedures as expressed in the logic of the time sequences presented in figure 7, and argued in favour of a flexibility-stability dilemma approach to

satisfucing

---

17. Please notice that the values of the curve which describes learning are greater than zero, both initially and at the end of the process.

the analysis of the interaction in question. However, the nature of that interaction was, to a considerable degree, confined within a black box for the reason of illustrative simplicity.

Now, following the approach of Hage & Aiken (1970) and Zaltman, Duncan & Holbek (1973), the black box is about to become illuminated. According to this approach, the process displayed in the simple model may be described as iterative and based on the overlap between the stages of initiation and implementation[18], which are depicted in figure 8 where the equivalent concepts of the two monographs are placed at the same horisontal levels.[19] At the stage of initiation, the organisation members become aware of the existence of a solution to performance gaps, and this solution is subjected to a decision making process where it is decided wether to adopt or reject the solution in question.[20] At the stage of implementation, the first attempt to utilise the solution in question is made, and a final implementation occurs through trial-and-error. Performance gaps are defined as

could do + what it does

> discrepancies between what the organization could do by virtue of a goal-related opportunity in its environment and what it actually does in terms of exploiting that opportunity. [...] When a discrepancy exists between what the organization is doing and what its decision makers believe it ought to be doing, there is a performance gap. [...] Given the assumption that the organization is aware of the gap, a need can be said to exist. This involves another assumption: the performance gap is percieved as having significant adverse consequences for the organization if the gap is not narrowed or bridged. The awareness and need, in effect, *unfreeze* elements within the organization most closely related to the external environmental change. When this occurs, conditions are present for altering the structure and function of the organization or some subsystem of it.
>
> (Zaltman et al., 1973, pp.2; 55; 3. Italics added)

performance gap

The *initiation* stage in the Zaltman et al. (1973) version comprises three substages:

---

18. Or search and decision, as parts of a decision making process.

19. Zaltman et al. (1973, p.182) differentiate themselves from Hage & Aiken (1970) from the point of view that Hage & Aiken (1970) do not "identify the organizational dilemmas". However, as it appears from figure 9 and the discussion related to figure 9, I am not fully in agreement with this point of view, although it must be recognised that Zaltman et al. (1973) place considerable more emphasis on feedback processess than Hage & Aiken (1970).

20. Of course, there is always the question whether a need stimulates the search for solutions, or whether the awareness of a solution stimulates the need for its implementation (Zaltman et al., 1973, p.62). Indeed, within an organisation there may exist, at any point in time, a number of solutions "looking for a problem", cf. the garbage can approach of Cohen, March & Olsen (1976) which is described in chapter 6.

(1) At the stage of *knowledge-awareness* a performance is either observed and triggers a process of search, or is percieved as the outcome of previous search since search "may alter the frame of reference held by decision makers" which "could lead to the perception of a performance gap, whereas prior to awareness there was no gap" (ibid., p.63). At this substage, organisational tension relates to the conflict between current organisational goals and the formation of new goals at the level of strategy.

(2) At the substage of *formation of attitudes* towards innovation, the organisation member respond to the innovation proposal according to how open they are to changes of performance programs and how they percieve the potentials of the organisation for innovation. At this substage, organisational tension relates to the relationship between the attitudes of subordinates and the attitudes of those in charge.

conflict emerges from changing goals

Figure 8. *Stages in the innovation model of the flexibility-stability approach*

Hage & Aiken (1970)	Zaltman et al. (1973)
Evaluation	Initiation stage
	- knowledge-awareness
	- formation of attitudes
Initiation	- decision
Implementation Routinisation	Implementation stage - initial implementation - continued-sustained implementation

Source: Zaltman et al. (1973, pp.181-82) & Hage & Aiken (1970, pp.94-106)

(3) At the substage of *decision*, the proposals for changes of performance programs are evaluated and a decision is made. At this substage, organisational tension relates to on what terms the information about feasible solutions should be evaluated, and thus to where the organisation finds itself in relation to the probabilities of positive *vis-á-vis* negative outcomes of the change decision.[21]

The *implementation* stage in the Zaltman et al. (1973) version comprises two substages: (1) At the stage of *initial implementation*, "the organization makes the first attempt to utilize the particular innovation" (ibid., p.67). At this substage, organisational tension

---

21. That is, evaluation in the sense of March & Simon (1958), see figure 16 in chapter 3.

reminds of the tensions that may be observed at the decision substage, although the acquisition and processing of knowledge is more of a trial-and-error nature. Finally, (2) at the substage of *continued-sustained implementation* the organisation commits itself to the innovation in question in the sense that the change of performance programs is allowed to generate, and is based on, a change of operating procedures. *different behav*

Despite the existence of feedback loops, the processess of initiation and implementation are not easily made compatible, not only due to various organisational tensions as mentioned above, but also because they may require different kinds of behaviour of the organisation members.[22] For instance, while the initiation of some new sales-production coordination mechanism may require that a number of organisation members transcend the cognitive borders of their subunits or departments as part of a cooperative process, the implementation may require a relatively unambiguos definition of the division of labour between different kinds of organisational roles.[23] A situation like this implies that different types of intraorganisational rationalities should be integrated, both vertically and horisontally, at least to the extent where some minimum performance level is met, and the process of integration is likely to involve changes in operating procedures, as indicated by the simple model of section 2.1. The change of operating procedures involves the change of performance programs, i.e. the way in which the organisation respond to alterations of the task environment (cf. March & Simon, 1958, p.141-42).

Now, the degree to which performance programs are changed is likely to vary according to the nature of integration of intraorganisational rationalities. Based on seven organisational characteristics that denote the type of integration, figure 9 outlines some general propositions about the relationship between the integration of intraorganisational rationalities and the rate of change of performance programs. The analytical focus of figure 9 is sociological and describes the organisation design in terms of variables which reflect the diversity of skills, the degree of routinisation of jobs, the intraorganisational distribution of power and rewards, and organisational goals in terms of efficiency[24] and

organic vs mechanistic

----

22. And thus different kinds of organisation design.

23. To some extent, the difference between the requirements posed to the organisation design by the stages of initiation and implementation, respectively, may be compared to the distinction between organic and mechanistic forms made by Burns & Stalker (1961), to whom the organisation design reflects how organisations cope with the nature of the task environment.

24. Efficiency is a broad concept which may describe the degree by which a set of goals may be satisfied or the degree by which effectiveness is obtained or emphasised within the organisation. The word "efficiency" denotes the ability to perform duties well, while "effectiveness" refers to the ability to accomplish predefined goals. Thus, in their work on the relationship between routine

job satisfaction. The relationship between these social variables and the rate of change of performance programs indicates that the outcome of learning depends on the organisation design, i.e. learning may be stimulated or hampered by the organisational characteristics for a number of reasons, some of which are described below.[25]

Organization characteristics that affect learning

Figure 9. *The dependence of the change of performance programs on organisation design*

disagree

Characteristics	Definition	Proposition
Complexity	Number of occupational specialities, and the degree of professionalism	The greater the complexity, the greater the rate of change of performance programs
Centralisation	Concentration of power and decision making	The higher the centralisation, the lower the rate of change
Formalisation	Degree of codification of jobs (routinisation)	The greater the formalisation, the lower the rate of change
Stratification	Distribution of rewards	The greater the stratification, the lower the rate of change
Production volume	Emphasis on quantity rather than quality	The higher the volume, the lower the rate of change
Effectiveness	Emphasis on the reduction of costs	The greater the emphasis, the lower the rate of change
Job satisfaction	The degree of morale among job occupants	The higher the degree of job satisfaction, the greater the rate of change

Source: Hage & Aiken (1970), pp.32-55

---

technology, social structure and organisational goals, Hage & Aiken (1969) distinguish between effectiveness and efficiency (ibid., p.374).

25. Zaltman et al. (1973, pp.178-81) provide an overview of the Hage & Aiken theory depicted in figure 9. In their own work, Zaltman et al. (1973) focus on complexity, formalisation and centralisation, cf. also Holbek (1988). The following description of these three variables draws on Gjerding (1992, pp.100-01), and an interpretation of the social variables in terms of various models for organisational learning is made in later chapters, cf. section 1.5.

diversity enlarges the # of sign posts fn action

Complexity refers to that aspect of the accumulation of knowledge in society that is reflected in the increasing number of occupations (Hage & Aiken, 1970, p.16), and the level of organisational complexity may be described in terms of the number of occupational specialties and the degree of professionalism in each specialty (ibid., p.33-35). A great number of occupations in the organisation creates a diversity of values and perspectives that serve as an impetus for change, because diversity enlarges the set of signposts for action and thus makes the organisation more receptive to opportunities for technical and organisational change. For instance, Hage & Aiken (1970) argue that any occupation will try to demonstrate the necessity of its organisational role by seeking new ways to improve organisational performance. Furthermore, a high degree of professionalism facilitates learning, because professionalism entails an emphasis on the acquisition of knowledge that induces the occupational members to keep abreast with the latest development in their field. Thus, diversity, or complexity in the sociological vocabulary, is conducive to learning. *diversity is conducive to learning*

However, diversity may, on the other hand, create a number of conflicts. Code scheme barriers may exist, e.g. interface problems between marketing and production concerning a new product as described in the simple model of section 1.1. Each subunit possesses its own focus of attention such as marketability in the case of the marketing unit and manufacturability in the case of the production unit. Furthermore, the impact of a change of performance programs on the social relations in the organisation might be regarded as a threat to the established structure of power and hierarchy, leading to resistance to change. Especially, a not-invented syndrome could be at work, e.g. because the subunit members may percieve the organisational configuration of that unit as unique for attaining a specific set of goals, and this perception leads to "the belief that alteration in the organization would dissipate this uniqueness" (Zaltman et al., 1973, p.87).

Formalisation and *centralisation* refer to, respectively, the "emphasis placed within the organization on following specific rules and procedures" (Zaltman et al., 1973, p.138), and "the locus of the authority and decision making in the organization" (ibid., p.143). While the variable of formalisation describes the extent to which the behaviour of the organisation members is codified and thus confined within a set of rules, the variable of centralisation describes the structure of power relations within which the rule-guided behaviour takes place. Power is, of course, only one way of creating organisational integration, and should, furthermore, not be interpreted as only visible power, i.e. the type of organisational behaviour regulated by explicit orders, but also as invisible power, i.e.

following specific rules & procedures
code scheme barriers

subunit focus of attention
resistance to alteration of organisation
formalised + centralised

53

the type of organisational behaviour that is regulated by some sort of social standard. Without going into details at the present moment[26], power may be defined simply as

> the capacity of one social position to set the conditions under which other social positions must perform, that is, the capacity of one social position to determine the actions of other social positions.
> *visible + invisible power* (Hage & Aiken, 1970, p.19)

Now, since the existence of rules "set limits not only on what men do but also on what men think" (ibid., p.43), one may argue that a high degree of codification limits the set of signposts for action and consequently the flow of proposals regarding new solutions and their adjustment during the stage of implementation. The functional role of rules is to secure uniformity in organisational behaviour in order to increase the predictability and decrease the uncertainty of organisational behaviour, and this may hamper the rate of change of performance programs since rules are often applied by decision makers on the assumption that the routinised behaviour in question represents some sort of best-practice for enacting a certain behavioural response. The organisation members are required to subject themselves to the behavioural patterns imposed by the rules, and in the case of a performance gap the individual member may be confronted with conflicting role expectations since he is supposed to change the behaviour that causes the performance gap while at the same time stick to established rules and procedures (Zaltman et al., 1973, p.139). In a situation where the nature of a performance gap requires a substantial change of performance programs, a low degree of formalisation might be needed to initiate and implement solutions, but in order to avoid role conflicts and organisational tension a "singleness of purpose is required" (ibid., p.140). While a singleness of purpose seems especially important in the last stages of the change of performance programs, a high degree of formalisation and centralisation limit the number of communication channels and restricts the number of avenues for feedback.

In consequence, decision makers may never be informed about occurring problems. Zaltman et al. (1973, p.143) elaborate on this point:

> A strict emphasis on hierarchy of authority often causes decision unit members to adhere to specific channels of communication and selectively to feed back only positive information regarding their job.
> Zaltman et al. (1973, p.143)

problems w/ Fordism

---

26. The concept of social standard refers to a broad range of phenomena such as the social values, the impact of artifacts on individual behaviour, and the basic assumptions about human life held by the human agents (e.g. Schein, 1985).

high degree of codification limits the sign post for action

At the same time, in some cases management do not want to be informed about performance gaps at all. For instance, Hage & Aiken (1970, pp.38-40) invoke the iron law of hierarchy arguing that those in power seek to preserve their power. In cases where the solution to a performance gap may threaten the hierarchial structure of the organisation, the change of performance programs may be vetoed by managers or dominant coalitions of organisation members.[27]   *iron law of hierarchy*

Stratification and *job satisfaction* refer to, respectively, the vertical job ladder and the pleasure that each individual finds in the job he is doing. Traditionally, the existence of a number of social, i.e. occupational, strata within an organisation has been percieved as a means of motivating a work force faced with the opportunities of climbing the job ladder. However, although a stratification system may stimulate job performance by providing a "clear line of promotion with ever-increasing rewards", it may, at the same time, discourage "suggestions for change because implicitly, if not explicitly, criticisms of present arrangements are criticisms of those who instituted them" (Hage & Aiken, 1970, p.46). In consequence, vertical communication may be restricted to an extent which negates performance gaps, and horisontal communication may be restricted as well, especially in cases where organisation members compete for rewards. While a strict emphasis on hierarchy of authority may induce the organisation members only to feed back positive information on their job performance, as argued above, the same phenomenon may occur as a consequence of incentive structures if organisational rewards are associated only with performance levels.   *horisontal communication*

Regarding job satisfaction, a high level of satisfaction may clearly be negatively associated with resistance to change and positively associated with the rate of change of performance gaps, since the organisation members may be more committed to the organisation and thus "more receptive to ideas for improving existing products or services" (ibid., p.53). However, to the extent that job satisfaction is related to the position of the organisation member in the stratification system, the relationship between satisfaction and resistance may be positive as a result of the dysfunctions of the incentive structure just described. In sum, one may argue that the relationship between job satisfaction and learning in terms of change of performance programs rely on the relationship between job satisfaction and stratification, which may lead to various types of breakdowns of the feedbacks between the stage of initiation and implementation.

dysfunctional incentive structure

---

27. Chapter 6 argues that these instances of diverted organisational learning may be described by the concepts of role-constrained, audience, and superstitious learning (March & Olsen, 1976, 1976a).

This discussion may be compared to the discussion of what motivates organisation members to concieve ideas for improvement, i.e. the answer to the following question: What is the relationship between motivation and individual creativity? This issue is often discussed with reference to the attitudes of the organisation members towards the task that they have to perform and their perceptions of what motivates them to actually undertake the task. As argued by Amabile (1988), task motivation may be *extrinsic*, i.e. motivation is achieved through external control of how the organisation member perform his task, or *intrinsic*, i.e. motivation comes as part of the task itself, for instance as an individual satisfaction with certain achievements. Amabile (1988) reviews a part of the body of literature on this subject, and her review seems to suggest that intrinsic motivation is more important to individual creativity than extrinsic motivation, and that "extrinsic constraints in the work environment can indeed undermine creative performance" (ibid., p.154).

Finally, *production volume* and *effectiveness* refer to the emphasis placed on performance in terms of quantities and the minimisation of costs. If the set of organisational goals emphasises a continuous production flow, any change that may interrupt this flow is likely to meet resistance, especially when the production volume is high. As Hage & Aiken (1970) argue:

intrinsic = ↑ creativity

> To paraphrase an old cliché, nothing fails like success. The organization with a high volume of production is likely to be resistant to the development of new products or services. The rationale used by organizational elite is that if production is high, everything is functioning adequately. Under these circumstances, it is less likely that anyone will raise a question whether improvement is necessary.
>
> (Hage & Aiken, 1970, p.49)

money

Furthermore, an emphasis on efficiency, measured as unit costs, may discourage innovation, since changes of performance programs are often costly and unpredictable, and involve a trade-off between types of costs, for instance a reduction in labour costs achieved through the spending of "exorbitant sums of money on machines and space" (Hage & Aiken, 1970, p.51), as in the case of traditional automation.

Summing up, the relationships between the organisational characteristics and the rate of change of performance programs depends on the existence of feedback mechanisms and the behaviour of the organisation members, both of which is influenced by the decision style of the organisation in question. At the initiation stage, feedback occurs as part of an iterative process where the acquisition and processing of knowledge change the agenda of search as information about innovation opportunities disseminates through the organisation. At the implementation stage, feedback occurs through trial-and-error and by way of change of the task environment caused by increased used of the innovation in question. Feedback may occur in relation to the evaluation of the organisational context, the assessment of possible solutions to performance gaps, the working of the innovation itself, and the ability of the change of performance programs to solve the performance gap in question.[28] The nature of these feedback mechanisms is sensitive to the interaction between the social and the technical subsystems, as argued above, since

> technical deficiencies cannot always be adequately explained and coped with unless feedback from the social system is established and corresponding information is interpreted. Simarily, emerging deficiencies related to the social system often require feedback from the technical system.
>
> technology   feedback   (Zaltman et al., 1973, p.75)

The decision style of the organisation may be described in terms of the organisational characteristics which determine the occupational diversity, the delegation of authority, and the degree of codification, i.e. the variables of complexity, centralisation and formalisation. According to figure 9, during the stage of *initiation*, the rate of proposals for the change of performance gaps is positively related to a high degree of complexity and a low degree of formalisation and centralisation, while the rate of change of performance gaps during the stage of *implementation* is positively related to a low degree of complexity and a high degree of formalisation and centralisation, both of which contribute to a "singleness of purpose". Thus, in order to solve the flexibility-stability problem, a mix of organisation design as suggested in figure 10 seems adequate.

Figure 10. *Mixing a solution to the flexibility-stability dilemma*

Innovation stage	Complexity	Formalisation	Centralisation
Initiation	High	Low	Low
Implementation	Low	High	High

Source: Zaltman et al. (1973), pp.134-46

Centrally controlled organisation = difficult to get something ww going

28. Zaltman et al. (1973, p.74) use, respectively, the terms context, input, process, and product evaluation as they refer to Stufflebaum (1967).

The pattern of organisational design revealed by figure 10 is one of *differentiation* between the stages of the innovation process, which may take place in time or space (Holbek, 1988). Differentiation in *time* takes place when the organisation "shift its structure while moving through the various stages of the innovation process" (ibid., p.255), and differentiation in *space* takes place when the organisation locates the stages of initiation and implementation to different parts of the organisation. As described by Holbek (1988, pp.260-62), differentiation in space may be horisontal or vertical. In the horisontal case, the innovation is moved between subunits according to the stage of innovation, e.g. initiation may take place in the R&D department or a project group, while implementation takes place in production and marketing. In the vertical case, initiation is often undertaken at the level of top management, while the task of implementation is transferred downwards, for instance by way of dictum. However, vertical differentiation is bound to involve horisontal differentiation, especially since most innovations have to be developed by certain parts of the organisation. Consequently, horisontal linkages are salient features of differentiation in space.

horizontal
+ vertical

2.3. Some critical remarks

The arguments on the organisational characteristics presented in figure 9 were derived from the sociological approach of Hage & Aiken (1970) and Zaltman et al. (1973). The approach was termed "classic" in the sense that it represents an original contribution at a point of time where the relationship between technical and organisational innovation was becoming a theoretical and empirical issue in a larger part of the scientific community. Of course, the approach should be understood and interpreted in terms of the industrial relations and conditions of that point in time. In a contemporary setting, the historic specificity of the classic flexibility-stability approach raises a number of questions.

sociological approach

First, the analytical and empirical background of the classic approach has been elaborated and revised during the last twenty years. The task of reviewing this development is outside the scope of the present study besides what will be discussed in relation to the neo-behavioural approach presented in chapter 6 and its application in chapter 7. The core of the neo-behavioural approach presented is the notion of incomplete cycles of learning, which elaborates on the classic flexibility-stability approach.

Second, during the last twenty years that have passed since the work of Hage & Aiken (1970) and Zaltman et al. (1973), the industrialised economies have experienced a shift in what Perez (1983, 1989) and Freeman & Perez (1988) have termed the *technoeconomic paradigm*. As will be argued in chapters 7-9, the shift of technoeconomic paradigm is associated with a change of the dominant general principles for intraorganisational

58

techno economic paradigm

coordination, and the types of flexibility-stability dilemmas which may occur have changed somewhat.

Third, although most of the discussion of the organisational characteristics presented in figure 9 seems to be valid in a contemporary setting, at least the variables of production volume and efficiency requires some qualification, particularly from the point of view that a shift in the technoeconomic paradigm is appearing contemporarily. Regarding the classic argument on the organisational emphasis on production *volume*, one may provide the counter-argument that although this perception made by the organisation members may be true in the short run, the persistence of performance gaps will tend to render it false in the long run. As empirical evidence suggests, the high-volume organisations which are successful in the long run are those able to institute technical and organisational changes (cf. e.g. Chandler, 1992).[29] Regarding the classic argument on the organisational emphasis on *efficiency*, it should be noted that Hage & Aiken (1970, p.51) assume that firms often overlook that innovations are "spending more of one resource in order to save another". Although this trade-off may be found in a number of cases, the evidence on long run technical and organisational change seems to indicate that the trade-offs should be interpreted in relative terms, for instance as increasing capital costs relative to labour costs, but decreasing capital and labour costs relative to product prices. Freeman & Perez (1988) argue that in the long run technical and organisational change involve a decrease of the absolute and relative prices of certain key technology factors, and the evidence on advanced manufacturing technology indicate that innovation may be both labour-, capital-, material- and energy-saving (e.g. Jelinek & Goldhar, 1983; Sundqvist et al., 1988; Zairi, 1992), however in proportions depending on the actual style of implementation (e.g. Haywood & Bessant, 1987). Furthermore, the advent of advanced manufacturing technology has lead to a continual downward push of the trade-off between production scale and costs effectiveness (Bessant, 1989). Chapter 8 return to these issues.

Associated with the issue of the historical specificity of the flexibility-stability approach, it should be stressed that the theoretical argument in section 2.2 foused on *intra*organisational arrangements. However, according to contemporary empirical evidence, one may argue that differentiation in space should also be treated as an *inter*organisational phenomenon in the sense that the initiation stage is located in one organisation and the implementation stage in another. A number of interorganisational arrangements are becoming increasingly frequent, such as joint-ventures, subcontracting,

---

29. Furthermore, the advent of microelectronics and information technology has changed the issue of production volume and efficiency, as will be argued in chapters 7-8.

licensing and leasing arrangements, and venture management, both internal and external, is to an increasing extent utilised as a means of overcoming organisational inertia.[30]

Regarding differentiation as a solution to the flexibility-stability dilemma, some additional problems of organisation design may come to mind.

First, if we contemplate the issue of uncertainty in terms of the ability of the organisation to attach probabilities to preferred outcome, vertical differentiation of space seems likely to entail the *paradox of administration* described by Thompson (1967).

On the one hand, an organisation may try to seal off its technical core, i.e. production activities, from contingencies creating uncertainty in order to secure smooth operation; on the other hand, smooth operation basically calls for a computational decision strategy which requires that preferable outcomes and cause-effect relationships have been formulated at the levels of tactical and strategic decision making. However, organisational goals are, essentially, characterised by diversity since they have to deal with a number of contingencies. Consequently, in *poradox of adm*

> the *short run*, administration seeks the reduction or elimination of uncertainty in order to score well on assessments of technical rationality. In the *long run*, however, we would expect administration to strive for flexibility through freedom from commitment - i.e. slack - for the larger the fund of uncommitted capacities, the greater the organization's assurance of self-control in an uncertain future.
>
> (Thompson, 1967, p.150)

These considerations leads Thompson (1967, pp.151-52) to elaborate on the style of search behaviour invoked by the behavioural theory of the firm, an issue further discussed in chapter 3, and he suggests that an organisation may apply a strategy of search that is equivalent to the curiosity of the individual, i.e. *opportunistic surveillance*:

> It is possible to concieve of monitoring behavior which scans the environment for opportunities - which does not wait to be activated by a problem and which does not therefore stop when a problem solution has been found.
>
> (Thompson, 1967, p.151)

This type of behaviour corresponds to the phenomena of solutions searching for a problem (Cohen, March & Olsen, 1976) mentioned earlier, and it corresponds to the observation of Zaltman et al. (1973, p.62) that the need for the implementation of a change of

---

30. Cf. for instance Littler & Sweeting (1984), Burgelman & Sayles (1986), Mowery (1989) and Hagedoorn (1990). In a Danish context, Jensen (1992) describes the cooperation between producers in the electronics industry, and Christensen, Andersson & Blenker (1992) describe the various types of cooperative relationships between industrial producers and suppliers.

performance programs may be stimulated by the awareness of a solution. To the extent that these mechanisms are at work, the simple model of section 2.1 may describe proactive as well as reactive behaviour, because performance gaps in the former case appear as perceptions of change opportunities which may enhance the performance of the organisation, although the present level of performance meets the minimum requirements (cf. Zaltman et al., 1973, pp.2-3; 55).

Second, following the reasoning of the simple model in section 2.1, horisontal differentiation in space is likely to involve the conflict between lower-level rationalities, especially since different types of rationalities may enter the innovation process at different points in time.

For instance, in developing a new product the R&D department may be concerned with the incorporation of the latest technical advance and pay less attention to the technical capabilities of the production unit. The production department, on the other hand, may focus on product improvements that utilise the existing technical capabilities. If communication between these two groups is infrequent and to some extent punctuated, the ensuing innovation proposals from the R&D department usually will pay less attention to manufacturability than to technical sophistication, and in consequence the innovation may be opposed by the production unit. We may elaborate on the scenario and argue that the conflict becomes tense, if for instance the sales & marketing department judges the sales prospects of the new product differently from the R&D and production departments. To the extent that communication between the R&D and sales & marketing departments is inefficient, e.g. due to large differences in perceptions and priorities, the introduction at the market could fail despite of ensuing adjustments undertaken by the production department, for instance because the sales & marketing department fails to appreciate the functionalities of the product or because the R&D department fails to appreciate the needs of the market. These types of interface problems are adressed throughout the study.

Third, from an intraorganisational point of view, pure differentiation in time is difficult to comprehend, unless we imagine a number of special cases.

One case is the growth of a firm which is founded on an innovative idea, develops it, produces and markets it, and becomes increasingly bigger in the process. This is a type of organisational life cycle well-known from the industries of consumer electronics. Another case is the variability of a small firm where the organisational configuration tends to vary over the life cycle of the innovation. This case is most likely to be found in customised production such as the development of computer programmes for specific purposes. Most likely, intraorganisational differentiation in time is only found in combination with differentiation in space, e.g. in the case of a (larger) organisation where a number of organisation members are able to participate in both organic and mechanistic forms that may be combined through parallel organisation (Holbek, 1988, pp.263-64).

Chapter 3

A Behavioural Theory of Organisational Action

3.1. The appearance of the behavioural theory of the firm

The behavioural theory of the firm (Cyert & March, 1963) appeared in opposition to the assumptions of mainstream economics, primarily directed towards the assumptions of rational decision making associated with the concept of economic man (Simon, 1957). The nature of rationality invoked by mainstream economics is *substantive* in the sense that

> the rationality of behavior depends upon the actor in only a single respect - his goals. Given these goals, the rational behavior is determined entirely by the characteristics of the environment in which it takes place.
>
> (Simon, 1976, p.130-31)

In consequence, the behavioural theory also challenged the *á priori* status of neoclassical theorising. The standard neoclassical case holds the vision of an economic system with fully-informed economic agents free to allocate productive resources among alternative uses in response to price changes, but small enough to lack discretionary power. The overall rationality may be Walrasian, where the economic agents adjust prices to given quantities, or Marshallian where the agents adjust quantities to given prices. Whichever adjustment mechanism is at work, the behaviour of the economic agents is determined by the characteristics of the economic environment, which forces them to maximise profits or experience extinction. Thus, the neoclassical firm appears as a theoretical construct with an *á priori* character objectively determined by the assumptions of the competitive forces in the external environment, and "all of the empirical content of this neoclassical model lies in the description of the environment" (Cyert & Hedrick, 1972, pp.398-99).

One may argue that the behavioural theory appeared as part of an increasing heterodoxy of the theory of the firm.[31] Originally, the *post-Keynesian* theory of the firm developed in opposition to the neoclassical conception of the entrepreneur as a reactive profit-maximising agent subject to the conditions of perfect competition. Instead, it advocates the principle of full-cost pricing (Hall & Hitch, 1939) as a general decision rule in combination with the idea that firms to some extent possess discretionary market power (Robinson, 1933; Chamberlin, 1933) and may set barriers to market entry (Bain, 1949;

---

31. The following, extremely brief, account is inspired by Marris & Mueller (1980) and Knudsen (1991).

1956). Parallel to this theoretical development, the *internal organisation* approach appeared, taking as its point of departure the notion of markets as being costly and imperfect (Coase, 1937) and the Simonian concept of bounded rationality (Simon, 1957). The core of the internal organisation approach is a focus on the need for internal organisation in order to make economic activity socially functional and productive, e.g. in cases of the shirking of teams (Alchian & Demsetz, 1972) and asset-specificity and opportunistic behaviour (Williamson, 1975). Interacting with the internal organisation approach, the *managerial approach* evolved, preoccupying itself with the the motives of the economic agents, which are assumed to be guided by a utility function maximised subject to a minimum profit constraint (Baumol, 1959; Marris, 1964; Williamson, 1964). The act of maximisation is by no means trivial, partly due to the existence of bounded rationality, and the maximisation of utility function hypothesis is relaxed in the *behavioural theory* which employs the concept of satisficing behaviour.[32]

Satisficing behaviour

3.2. Satisficing behaviour and bounded rationality

Satisficing behaviour and bounded rationality are intertwined concepts. They reflect *procedural rationality* in the sense that the rationality of behaviour "depends on the process that generated it" (Simon, 1976, p.131). Satisficing behaviour implies that the alternatives, on which the economic agent contemplates, are discovered sequentially through search processes guided by action programs. These behavioural programs determine which types of alternatives are invoked or searched for, and the alternatives chosen are those that meet or exceed some minimum criteria (March & Simon, 1958, pp.140; 169). This type of decision making behaviour occurs, because the economic agent recognises *behaviour depend on the process that generated it.*

that the world he percieves is a drastically simplified model of the buzzing, blooming confusion that constitutes the real world. He is content with this gross simplification because he believes that the real world is almost empty - that most of the facts of the real

---

32. This type of organisational behaviour is also found in the fields of evolutionary economics (Nelson & Winter, 1974; 1982) and the institutional game-theoretical X-efficiency approach (Leibenstein, 1966). Even though the brief account made above may indicate that there has been a movement towards scientific progress in the sense that (1) the theoretical models approaches the reality of real-life firms or (2) that a type of Kuhnian normal science is being established, one can hardly argue that the hard core of mainstream economics have been affected. As argued by Marris & Mueller (1980), the survey presented by Cyert & Hedrick (1972) indicates the opposite, and later work done by Knudsen (1991) seems to confirm this view. This state of affairs may partly be explained by the large diversity of theoretical perspectives which create prohibitive barriers to interdisciplinary discussion and cross-fertilisation.

world have no great relevance to any particular situation he is facing, and that most significant chains of causes and consequences are short and simple.

(Simon, 1957, p.xxv)

The agent does not make comprehensive choices like the archetype of economic man, but concerns himself with rather specific matters that might not even be the ones important to his decision (Simon, 1983). He does not employ detailed scenarios of the future, but focus his attention on some specific values to the relative neglect of others, and the particular decision domain on which he operates will evoke some values, and not others. While a large part of his decision making process is devoted to the gathering of facts and the evoking of the values in question, "the choice itself may take very little time" (Simon, 1983, p.19). Thus, "the decision process may in fact be highly routinized" (March & Simon, 1958, p.148). *decisions may be routinized*

Satisficing decision making has a number of relative advantages compared to the alleged behaviour of economic man. It frees the economic agent from the impossible task of taking into account all facts of consequence to the choice in question and limits the costs of search by economising on the gap between the "real environment of a decision" and "the environment as the actors percieve it" (Simon, 1978, p.8). However, if one percieves the organisation as an adaptive coalition (Cyert & March, 1963) where the stability of the set of goals is essential in order to keep the coalition together, the relative advantage of satisficing behaviour may become a disadvantage, because there is a tendency towards dormancy[33] which occurs for two reasons: *Search & satisficing*

First, the search process associated with satisficing behaviour terminates when the minimum criteria are met or exceeded. These criteria is determined by the aspiration level of the economic agent. However, the aspiration level itself "adjusts gradually to the value of the offers recieved so far" (Simon, 1978, p.10), and to the extent that the aspiration level depends on the stability of the set of goals, there is a tendency towards fixation of aspiration levels. The tendency towards fixation of aspiration levels is associated with the segmentation of organisational goals and decision making, where the organisation in order to resolve conflicts within the coalition relies on *local rationality* in the sense that subunits "deal with a limited set of problems and a limited set of goals" (Cyert & March, 1963, p.117). *Second*, adaptive satisficing behaviour implies that the organisation seeks to avoid uncertainty by maintaining a set of simple rules (Cyert & March, 1963, p.102), often described as "rules of thumb". Search becomes *problemistic*, i.e. search is triggered by a problem and proceeds "on the basis of a simple model of causality until driven to a

search as problemistic.

---

33. As implied by the simple structural-functionalist model presented in section 2.1.

local = limited set
simple rules of thumb

65

more complex one" (ibid., p.121). This type of search is simple-minded and local in the sense that it takes place in the neighbourhood of problem symptoms and current alternatives, and search becomes distant only when local search fails.[34] In a behavioural setting, the application of local search is as an inevitable outcome of the application of local rationality in order to keep the coalition together, since distant search tends to stimulate the reallocation of resources and functions and thus disturb the stability of the set of goals.

The way in which the local rationalities are coordinated, and the stability of the set of goals is preserved, depends on the relationship between the organisation design and the awareness of the organisation member of the fact that goals may be conflicting. The organisation design may be described in terms of procedural and substantive coordination. *Procedural coordination* refers to the specification of the organisation, i.e. the lines of authority and responsibility, while *substantial coordination* refers to the contents of the organisational activities (Simon, 1957, p.140). Procedural and substantial coordination serve to focus the attention of the organisation member on specific activities and problems, the solution of which is guided by performance programs in order to ensure that the solutions "are consistent with a large number of other solutions to other tasks being performed in the organization" (Cyert & March, 1963, p.104). The focus of attention is an important element in the organisation member's identification with the objectives of the organisation (Simon, 1957, p.210) and explains

> why organizations are successful in surviving with a large set of unrationalized goals. They rarely see the conflicting objectives simultaneously.
> (Cyert & March, 1963, p.35)

design sets limits on member rationality

The focus of attention is determined by the way in which the organisation design sets limits on the rationality of the organisation member.[35] These limits are embedded in the intraorganisational task environment, which is determined by a number of circumstances

---

34. In that case, "we assume two developments. First, the organization uses increasingly complex ("distant") search; second, the organization introduces a third search rule: (3) search in organizationally vulnerable areas" (Cyert & March, 1963, p.122). The two search rules associated with simple-minded search are: "(1) search in the neighborhood of the problem symptom and (2) search in the neighborhood of the current alternative" (ibid., p.121).

35. To an important extent, the focus of attention depends on the individual motives and wants, mental capacities and social relations of the organisation member (Simon, 1957; March & Simon, 1958). However, although the analysis of these causes and effects is an intrinsic part of the behavioural theory of the firm, they will not be adressed in the present study, which is devoted to the analysis of collectivities of human beings within the organisation.

local is based on current realm of problems + alternatives

associated with the division of labour, the nature of coordination, and the nature of communication processes. Figure 11 represents an attempt to depict the propositions of the behavioural theory on the relationship between (1) the focus of attention and (2) the intraorganisational goals and coordination. At the same time, these elements of the organisational configuration will serve as a point of reference for the ensuing discussion in chapter 4 on contingency theory. The key elements of figure 11 are span of attention, tolerance for interdependence, communication, and the differentiation and persistence of subgoals. Of course, these key elements are intertwined. However, figure 11 reflects an attempt to segregate the constituent parts of goals and coordination in order to present the behavioural *core* arguments and propositions on organisation design, primarily as argued by March & Simon (1958).

Span of attention refers to the mental capacity of the organisation member which determines the narrowness of the focus of attention. To the extent that time pressure is involved, the span of attention decreases as selective perception increases.

Subgoals refer to the way in which the task environment of the organisation member is organised. As argued by Simon (1958, pp.45-78), rationality in organisational action is determined by the construction of a hierarchy of ends, i.e. a number of means-ends chains where lower-level ends become higher-level means.[36] The existence of subgoals is the outcome of the way in which problem-solving is simplified by the delegation of authority and responsibility. Delegation means that the organisational tasks are factored "into a number of nearly independent parts, so that each organizational unit handles one of these parts and can omit the others" (March & Simon, 1958, p.151). Consequently,

subunit & organizational goal

> one of the most frequent phenomena in an organization is the differentiation of subunit goals and the identification of individuals with the goals of the subunit, independently of the contribution of that goal to the organization as a whole.
>
> (Cyert & March, 1963, p.19)

March & Simon (1958, pp.152-53) argue that the predominance of subgoals over goals at the level of the organisation as a whole has a tendency to become reinforced through selective perception of problems. First, there is reinforcement which appears at the level of subunits as stimuli to the *persistence of subgoals*. The content of in-group communica-

---

36. Likewise, Simon (1958, p.32) argues that there is "no essential difference between a *purpose* and a *process*, but only a distinction of degree. A *process* is an activity whose immediate purpose is at a low level in the hierarchy of means and ends, while a *purpose* is a collection of activity whose orienting value or aim is at a high level in the means-ends hierarchy" (quotation marks in the original text substituted by italics).

← *collective thought & individual thought*

tion, i.e. the selective exchange of information and perceptions within the subunit, determines the focus of information, i.e. what type of problems and means-ends relationship the organisation member attends to. Second, there is reinforcement which appears at the level of the intraorganisational environment of the subunit as stimuli to the *differentiation of subgoals*. The differentiation of subgoals is influenced by the selectivity of information that the organisation member recieve, and there is a strong affinity between the selectivity of the information recieved and the task environment, i.e. the division of labour in the organisation.[37]

division of labor is task environment

Figure 11. *A behavioural determination of the focus of attention*

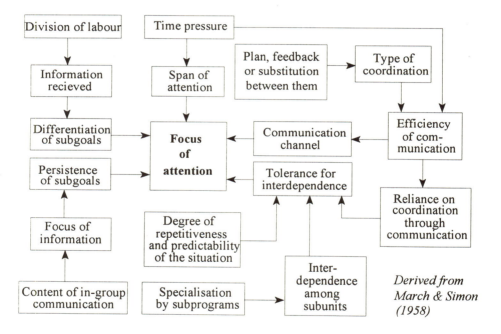

----

37. Third, March & Simon (1958, p.152) argue that there is reinforcement at the level of the individual "through selective perception and rationalization. That is, the persistence of subgoals is furthered by the focus of attention it helps to generate". In figure 11, this should be depicted by a double arrow between the differentiation of subgoals and the focus of attention. However, the same apply to all of the factors which in figure 11 are allowed to determine the focus of attention, and for the sake of simplicity these relationships have been omitted.

Communication refers to the way in which the organisation members relate to each other regarding procedural and substantial coordination, which according to March & Simon (1958, p.161) may be defined in terms of the occasions for communication. Communication regarding procedural coordination relates to the type of communication on new events which become neccesary when nonprogrammed activities occur, and the type of communication which is necessary to "inititate and establish programs, including day-to-day adjustment" (ibid.). Communication regarding substantial coordination relates to the type of communication which is necessary in order to provide data for and evoke activity programs, and evaluate the outcome. Communication run along communication channels which have a tendency to persist as they facilitate the interaction between organisation members. The set of communication channels is determined by the deliberate design of the organisation and the needs which evolve out of the daily problems and social interactions. Thus, in part   *Communication channels have tendency to persist*

> the communication network is planned; in part, it grows up, in response to the need for specific kinds of communication; in part, it develops in response to the social functions of communication.
>
> *channels deliberate + organic*   (March & Simon, 1958, p.168)

Communication channels reflect the formal and informal delegation of the types of problems and questions which the organisation member attends to. The usage of communication channels is reinforced for two reasons. First, if the efficiency of the communication channel is great, the channel tends to become used more frequently. Second, the efficiency of the channel is influenced by the dimension of time involved in the tasks on which communication takes place, in the sense that the established patterns of communication tend to become reinforced to the extent that time pressure is high.

Finally, the *tolerance for interdependence* refers to ability of the organisation to handle interdependencies among the constituent parts of the task environment. The interdependencies reflect the allocation of problems and answers through the construction of subprograms. To the extent that the execution of subprograms takes place within a stable task environment, the coordination of subprograms becomes relatively simple and the tolerance for interdependence relatively greater. Conversely, to the extent that the execution of subprograms is influenced by contingencies which are unanticipated or imperfectly predicted, the tolerance for interdependence becomes relatively smaller. The tolerance for interdependence is positively related to the efficiency of communication, because the propensity of the organisation to rely on coordination through communication tends to grow as communication becomes more efficient. The efficiency of communication depends on the type of coordination which reflects the nature of the contingencies on which the organisation member communicates. While coordination by plan occurs when the contingencies can be handled by pre-established programs, coordination by feedback occurs when the contingencies create a need to communicate on "deviations from planned or predicted conditions, or to give instructions for changes in activity to adjust to these deviations" (March & Simon, 1958, p.160). Communication tends to be more efficient in cases of coordination by plan than in cases of coordination by feedback where the number of sources for misinterpretations and deficient information increases. Consequently, communication tends to be more efficient in cases where feedback is substituted by plan. *coordination by feedback*

3.3. Restating the flexibility-stability approach

Now, restating the flexibility-stability dilemma in terms of the approach presented in figure 11, one must observe that the occurrence of performance gaps is associated with changes in the focus of attention.[38] Performance gaps, which occur at the focus of attention, reflect that (1) one or more organisation members have become aware of discrepancies between what the organisation is doing and what it might be doing in face of goal-related opportunities or perceptions of decision makers, and that (2) this awareness have been disseminated among the organisation members and decision makers. Changes in the focus of attention are related to changes of one or more of the key factors which affect the focus of attention, i.e. goals, communication, span of attention and tolerance for interdependence. The change of these key factors may, primarily, be explained in terms of the relationship between aspiration levels and the level of achievements, i.e. the ocassion for performance gaps to occur.

Previously, it was argued that aspiration levels adjust gradually to organisational performance, and that the existence of an organisational equilibrium creates a tendency towards the fixation of aspiration levels. Since *goals* reflect levels of aspiration, one should expect goals to be relatively stable. However, this argument does

Figure 12. *The relationship between levels of achievement and aspiration*

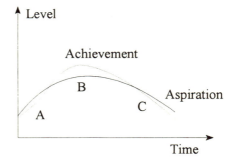

----

38. Please recall the defintion of performance gaps presented in section 2.2.

not imply that the organisational equilibrium is devoid of changes in the aspiration levels, as figure 12 suggests. The degree to which the aspiration levels tend to become fixed depends on the actual experience of the organisation, since the adjustment mechanism implies that the effect of the organisational equilibrium on the stability of the aspiration levels is mediated by performance. Following Cyert & March (1963, p.34), the level of aspiration tends to be a little bit higher than the level of achievement in periods of stability. This is the case of A where the level of achievement increases gradually. In the B case, the level of achievement increases progressively, and the aspiration level tends to lag behind. Finally, in the case of C, the level of achievement is decreasing progressively, and the aspiration level tends to become larger than the level of achievement.

This simple but intuitively appealing exposition implies that the focus of attention is bound to shift as the organisation encounters the extraorganisational task environment. The shifting of the focus of attention as the outcome of experience is, essentially, what the behavioural theory of the firm regards as *organisational learning* which becomes manifested in the adaptation of goals, attention rules and search rules (Cyert & March, 1963, pp.123-25). According to Cyert & March (1963), the adaptation of goals reflects experience in terms of the achievements made by the organisation in question and by other organisations by which the organisation in question compares its achievements. In the course of experience, attention rules are adapted since "organisations learn to attend to some criteria and ignore others" when achievements are evaluated, and to the extent that the organisation is sensitive to external comparisons, "organisations learn to pay attention to some parts of their environment and to ignore other parts" (ibid., p.124). As the outcome of this, search rules and the code for communication change in a way which tends to enforce a new regularity in the intraorganisational interaction. The dynamics of this tendency, which is comparable to the simple model in section 2.1, reflects that search and communication in the adaptive organisation are problem-oriented:

...when an organization discovers a solution to a problem by searching in a particular way, it will be more likely to search in that way in future problems of the same type; when an organization fails to find a solution by searching in a particular way, it will be less likely to search in that way in future problems of the same type. Thus, the order in which various alternative solutions to a problem are considered will change as the organization experiences success or failure with alternatives.
In a similar fashion, the code (or language) for communicating information about alternatives and their consequences adapt to experience. Any decision-making system develops codes for communicating information about the environment. Such a code partitions all possible states of the world into a relatively small number of classes of states. Learning consists in changes in the partitioning. In general, we assume the gradual development of an efficient code in terms of the decision rules currently in use. Thus, if a decision rule is designed to choose between two alternatives, the information code will

71

tend to reduce all possible states of the world to two classes. If the decision rules change, we assume a change in the information code, but only after a time lag reflecting the rate of learning.

(Cyert & March, 1963, pp.124-25)

satisficing

The type of decision making behaviour implied in the process of rational adaptation is, of course, satisficing in the sense described earlier, i.e. the amount of search depends on the degree of satisfaction, and search is selectively directed towards the neighbourhood of problem symptoms and current alternatives. As argued previously, there is an affinity between the importance of local rationality and the amount of local search, and to the extent that the extraorganisational task environment is stable and the use of acceptable-level decision rules yields satisfactory results, adaptation primarily takes place as local changes of performance programs (March & Simon, 1958, p.176). Thus, the image of an organisation presented by the behavioural theory of the firm is, primarily, the image of a goal-directed collective, the predominant characteristic of which is the persistence of the nature of organisational activities.

Innovation

In consequence, this theoretical perspective seems to leave little room for organisational and technical innovation. Innovation, which the behavioural theory define as "a new solution to a problem that currently faces the organization" (Cyert & March, 1963, p.278), becomes the exception rather than the rule and occurs primarily because the existing set of performance programs tends to yield unsatisfactory results. These occasions for innovation, which March & Simon (1958, p.184) describe as "natural" stimuli in the sense that they are not programmed, relate to the occurrence of performance necessities, optimum stress, and accidental opportunities. However, the behavioural theory does admit the possibility of innovation as a programmed activity which may become institutionalised through stimuli that supplement the natural stimuli to innovation (ibid., pp.184-85).

Innovation routinised

Figure 13. *A behavioural approach to the occasions for innovation*

Occasion	Characteristics of occasion
Necessity	Changes in the environment make existing programs unsatisfactory
Stress	Aspiration levels exceed achievement to a moderate extent
Opportunity	Accidental encounters not programmed
Programmed	Rates of change of performance are aimed for

Source: March & Simon (1958, pp.182-85) and Cyert & March (1963, pp.278-79)

Figure 13 provides an overview of the behavioural occasions for innovation. *Necessity*-stimulated innovation refers primarily to the case where the performance falls below some acceptable level due to changes in the extraorganisational task environment, and they represent the type of schocks in the adaptive rational systems model (Cyert & March, 1963, p.99) described in section 1.3. In the absence of schocks, innovation may, however, occur from within the organisation as stress-stimulated.

Stress-stimulated innovation appears as the outcome of experience in the sense that aspiration levels are positively related to achievements[39], however within a certain range defined by the level of optimal stress:

> ...if achievement too easily exceeds aspiration, apathy results; if aspiration is very much above achievement, frustration or desperation result, with consequent stereotypy. In the first case, there is no motivation for innovation; in the second case, neurotic reactions interfere with effective innovation. Optimal "stress" results when the carrot is just a *little* way ahead of the donkey - when aspirations exceed achievement by a small amount.
> (March & Simon, 1958, p.184)

— on the edge of chaos (handwritten)

Opportunity-related innovation refers to the case where the organisation is confronted accidentally with an opportunity. The term "accidentally" does not mean that opportunities occur by chance. They may be the outcome of a planned process which, however, lead to opportunities not anticipated in advance.

Finally, innovation may be a programmed activity. *Programmed*-stimulated innovation refers to the case where the organisation institutes innovation as part of the criteria of satisfaction, implicitly by stating the criteria in terms of rates of change or explicitly by stating some criteria in terms of rates of innovation. Programmed-stimulated innovation is problem-oriented like innovation stimulated by necessity, stress or opportunity, but is distinguished from the latter ocassions in that the problem is defined in advance.

A type of innovation different from the "natural" stimuli mentioned by March & Simon (1958), is *slack*-stimulated innovation. The difference may be attributed to two circumstances. *First*, Cyert & March (1963, p.279) argue that organisational slack, which they define as "the difference between the payments necessary to maintain the organization and the resources obtained from the environment by the coalition" (ibid., p.278), provides the resources for innovations that may have strong support from one or more of the constituent parts of the organisation, but may be rejected in the face of tight

---

39. Cf. figure 12. March & Simon (1958, p.183) argue that "the gradual upward movement of criteria will lead to periodic demands for innovation, but to only moderately vigorous effort". However, as implied by figure 12, such periodic demands may also occur as performance criteria move down.

budget constraints. Thus, slack-stimulated innovation differs from the types of innovation associated with necessity, stress and opportunity in the sense that (1) it does not, to the same extent, enjoy short-run legitimacy since it is not oriented towards pressing problems that is justifiable in the short run *vis-á-vis* a limited amount of resources; and (2) it rests upon organisational resources which are free to engage in search that is not directly related to pressing problems.

Second, like the other types of nonprogrammed innovation described above, slack-stimulated innovation fits easily into the adaptive rational systems model of the organisation. However, while the concept of innovation stimulated by necessity, stress or opportunity applies primarily to organisations which are unsuccessful in the sense that achievements fall below aspirations, slack-stimulated innovation is primarily applicable to organisations which are successful *per definitionem*, since only successful organisations are able to accumulate organisational slack. In consequence, the concept of slack-stimulated innovation allows the behavioural theory of the firm to explain why innovation occurs in cases other than incidents of adversity. The introduction of slack-stimulated innovation represents an important qualification to the behavioural theory of the firm, because the concept of satisficing behaviour in a model of adaptive rational behaviour tends to indicate that innovation only occurs as a reactive response to cases of adversity. As Cyert & March (1963, pp.278-79) are well aware, this is not the only reason why organisations innovate. To some extent, the concept of slack-stimulated innovation may be compared to the socalled Schumpeter hypothesis, that is the idea that innovation is most likely to occur in circumstances where firms are large, have none or few competitors, and thus are able to appropriate quasi-rents and accumulate resources for innovation. However, although there exists a number of empirical studies on the relationship between innovation and market power, the evidence seems inconclusive (Kamien & Schwartz, 1982; Nelson, 1987). Organisations may innovate because or despite of adversity, and this seems to validate the more balanced approach suggested by Cyert & March (1963) as opposed to the more restricted approach of March & Simon (1958).

From a behavioural point of view, one may argue that the incidence of slack-stimulated innovation tends to trade-off the influence of local rationality on the focus of attention.

March & Simon (1958, p.185) argue that the "propensity of organization members to engage in an activity" is highly influenced by the existence of time pressure and the clarity of goals. To the extent that time pressure is high, the organisation members tend to focus on the activity to which the time pressure is attached. To the extent that the activity is connected to goals which are explicit and/or easily understood by the organisation members, they tend to engage in this activity. Since the clarity of goals usually is higher in connection to the lower-level than the higher-level parts of the means-end hierarchy, and the organisation members more easily percieve time pressure in the local rather than the distant task environment, local rationality will tend to prevail over global rationality in relation to the engagement in activities.[40] However, organisational slack tends to provide some additional degrees of freedom not anticipated or created by the organisational design. While the sequential nature of the decision process on the one hand is determined by the set of activity programs and hierarchially ordered goals that determine the focus of attention, organisational slack does, on the other hand, enhance the opportunity of non-programmed activities that may result in the discovery of alternatives to performance gaps other than those invoked by programmed search. In consequence, slack-stimulated innovation may be positively correlated with both short-run and long-run adaptation, while the "natural" stimuli described by March & Simon (1958) may, on the other hand, be expected to be positively correlated with primarily short-run adaptation.

Characterising the process of innovation, March & Simon (1958) distinguish between reproductive and productive problem-solving, in both cases concepts that refer to the analytical level of the individual: *natural + slack stimulated innovation*

When problem-solving consists primarily in searching the memory in a relatively systematic fashion for solutions that are present there in nearly finished form, it is described as "reproductive". When the construction of new solutions out of more or less "raw" material is involved, the process is described as "productive".

(March & Simon, 1958, p.177)

productive + reproductive innovation processes

Reproductive problem-solving is mainly associated with programmed activity, while productive problem-solving is associated with nonprogrammed activity. Search associated with reproductive problem-solving involves the surveying of knowledge related to solutions in the neighbourhood of present activities, while search associated with productive problem-solving tends to become more distant. To the extent that slack-stimulated innovation is characterised by a long-run component, it is more likely to involve distant search and thus be a part of productive problem-solving.

Reproductive more local

---

40. An example of this, which appears in chapter 9, is the engagement in quality circles that is sensitive to the degree by which the objectives of quality management can be attached to the immediate tasks of the Q-circle members. The prevalence of local rationality over global rationality is the essence of *Gresham's Law* of planning, i.e.: "Daily routine drives out planning. Stated less cryptically, we predict that when an individual is faced with both highly programmed and highly unprogrammed tasks, the former tend to take precedence over the latter even in the absence of strong over-all time pressure" (March & Simon, 1958, p.185).

Figure 14. *Reproductive and productive problem-solving*

Degree of familiarity with solutions	Occasion for innovation	Search	Time
Reproductive problem-solving	Necessity Stress Opportunity	Local	Short-run
Productive problem-solving	Slack	Local and global	Long-run

Figure 14 summarises what has been said about the nonprogrammed occasions for innovation and reflects that reproductive and productive problem-solving may be associated with the occasions for innovation in terms of the extent to which they involve local or distant search. While local search is, primarily, associated with the occasions of necessity, stress and opportunity, distant search is likely to be involved in slack-stimulated innovation. However, the classification depicted in figure 14 represents an attempt to, *ideal-typically*, clarify the concept of innovation employed in the behavioural theory of the firm and should, accordingly, be interpreted with some care. Slack-stimulated innovation may *accidentally* result in favourable results, and thus we might expect that opportunity-stimulated and slack-stimulated innovation overlap in some instances. On the other hand, slack-stimulated and opportunity-stimulated innovation may be distinguished by their, respectively, long-run and short-run component. Furthermore, necessity-stimulated and stress-stimulated innovation are not likely to coincide with slack-stimulated innovation: Since organisational slack provides a buffer against shortcomings of achievements, necessity-stimulated innovation will not, *per definitionem*, coincide with slack-stimulated innovation; and since organisational slack reflects a state where the amount of resources obtained by the organisation is larger than the amount necessary to maintain the organisation, organisational slack is not likely to result in optimal stress. In conclusion, the classification suggested by figure 14 may provide a rather concise presentation of the behavioural occasions for innovation.

As has been described, reproductive and productive search are undertaken in order to overcome performance gaps. The results of the search process, i.e. the scanning or discovery of alternative solutions to a problem, are likely to cause a state of conflict in the decision situation depending on the degree of certainty involved in the evaluation of the results of search. According to March & Simon (1958), conflict may arise for three reasons, which can be located at a continuum of certainty, cf. figure 15.

In the case of *unacceptability*, the decision maker is able to assess the outcome of various alternative solutions, but even the most preferred one falls below the minimum level of satisfaction. In the case of *incompatibility*, the decision maker is able to predict the outcome of various alternative solutions, but cannot assess the value of these outcomes and hence is unable to prefer one solution to the other. Finally, in the case of *uncertainty*, even the outcome of various alternatives cannot be predicted. Figure 16 summarises the arguments of March & Simon (1958) regarding the occurrence of conflict in decision situations.

Figure 15. *Degree of certainty in the evaluation of alternative solutions to problems*

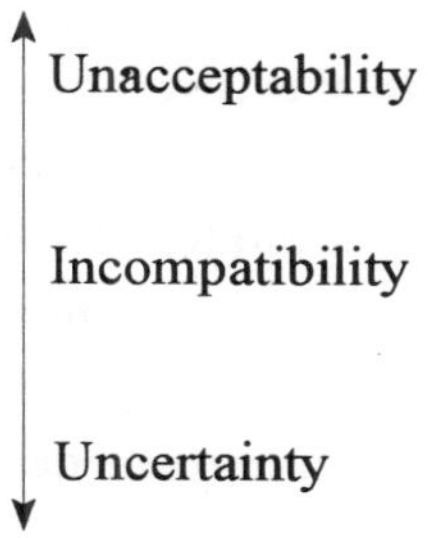

Unacceptability

Incompatibility

Uncertainty

Figure 16. *A behavioural typology of the choice between alternative solutions*

The probability of a positively valued state of affairs	The probability of a negatively valued state of affairs	The decision maker's assessment of the alternative
Large	Small	Good
Small	Small	Bland
Large	Large	Mixed
Small	Large	Poor
Unknown	Unknown	Uncertain

Source: Derived from March & Simon (1958, pp.113-14)

The behavioural core assumption on the nature of the decision situation is that there will be a motivation to reduce conflict. The strength of that motivation depends on the time pressure, which defines how quickly a solution must be found, and the availability of bland alternatives in the absence of good alternatives. To the extent that good alternatives are absent and time pressure is high, bland alternatives will be preferred as an escape hatch. In the normal case of *incompatibility*, the search rules described earlier implies that the process of decision depends on the sequence in which alternatives are considered, and the first alternative which supposedly meets the minimum criteria of satisfaction will be chosen. In consequence, the decision time tends to be relatively short (March & Simon, 1958, pp.116-17). In the case of *unacceptability*, the search process may be more comprehensive, and the conflict entailed in the decision situation depends more on the relationship between aspiration levels and achievement than on the insufficiency of knowledge. According to the dynamics discussed earlier in relation to the predominance of local rationality and the adjustment mechanism depicted in figure 12, the degree to which a solution is unacceptable depends on the time lag of adjustment of aspiration levels to achievements. Thus, the conflict entailed in the decision situation depends on the disparity between aspiration levels and achievements only to the extent that "the lag in that adjustment is appreciable" (ibid., p.120).

Finally, in the case of *uncertainty*, the conflict entailed in the decision situation is inclined to become tense, because there are, actually, no rules of the games which determine whether one solution is more appropriate than another. In this case, the rational systems model of adaptive organisations imply that two strategies may be feasible. Either the organisation must resort to the "third search rule" (cf. footnote 34) and trade off Gresham's Law of planning by applying distant search; this is more likely to occur in cases of organisational slack. Or the organisation may try to trade off uncertainty by relying on short-run adaptation, i.e. enact the present state of the focus of attention and sealing off those parts of the organisational activities which are the most vulnerable. Actually, Cyert & March (1963) argue:

> Organizations avoid uncertainty: (1) They avoid the requirement that they correctly anticipate events in the distant future by using decision rules emphasising short-run reaction to short-run feedback rather than anticipation of long-run uncertain events. (2) They avoid the requirement that they anticipate future reactions of other parts of their environment by arranging a negotiated environment. They impose plans, standard operating procedures, industry tradition, and uncertainty-absorbing contracts on that environment. In short, they achieve a reasonably manageable decision situation by avoiding planning where plan depend on predictions of uncertain future events and by emphasising planning where the plans can be made self-confirming through some control device.
>
> (Cyert & March, 1963, p.119)

The implications of the rational systems model of short-run, uncertainty-avoiding organisational behaviour for the flexibility-stability dilemma relate to two dimensions:

First, the implications to be inferred in relation to the flexibility-stability propositions outlined in figure 9 reflect that innovation in a behavioural setting is mainly of the routine type than the non-routine type. Furthermore, they reflect that the rational systems model of adaptive behaviour is focussed on the equilibrium of the organisation in the sense of Simon (1957)[41]: Organisational equilibrium, defined as a balance between the contributions from and the payments to the participants in the organisation, is the condition for the survival of the organisation. The organisational equilibrium depends on the ability of the organisational set-up to secure (1) a type of coordination of activities which allows the exercise of organisational authority to fall within the area of acceptance of the organisation members[42], and (2) a set of inducements able to bring about the type of activities necessary to sustain organisational efficiency.[43] Figure 17 summarises the behavioural implications for the flexibility-stability dilemma.

Second, as pointed out by Zaltman et al. (1973, p.168-69), the behavioural theory focusses on the stage of initiation, i.e. the stimuli which enact search rules, but not on the stage of implementation, i.e. the processes that occur when search is completed.[44] In my eyes, this may be explained by an over-reliance of the behavioural theory on the intraorganisational ability to enforce overall rationality by quasi-resolution of conflicts, which tend to establish an equilibrium between the various types of local rationalities. From this point of view, the organisational equilibrium, even though dynamic, is more sensitive to the processes that lead to a decision than to the processes involved in the implementation of that decision. Thus, the focus is on the elaboration of performance programs as a *process of decision* to adopt and institute rather than on the *process of instituting* the new performance programs.

Now, regarding the diversity of perspectives, which Hage & Aiken (1970) and Zaltman et al. (1973) associate with *complexity*, the behavioural theory tends to focus less on the number of occupational specialties and the degree of professionalism in each and more on the number of attention centres. An attention centre is an individual organisation

---

41. Who is inspired by Barnard's *The Functions of the Executive* (1938).

42. Simon (1957, p.133) defines the area of acceptance as the range of behaviour "within which the subordinate is willing to accept the decisions made for him by his superior". Barnard uses the term "zone of indifference" (ibid.).

43. Actually, March & Simon (1958, p.84) characterise the Barnard-Simon theory as "essentially a theory of motivation".

44. This is merely stated as a kind of fact by Zaltman et al. (1973) without further qualification than the reference to the focus on stimuli for search.

member or a collection of members, i.e. a subunit or an aggregate of subunits.[45] Because group problem-solving is superior to individual problem-solving regarding the amount of information which can be processed, the rate of change of performance programs is positively related to the extension of group problem-solving.

Figure 17. *A behavioural approach to the flexibility-stability dilemma, cf. figure 9*

Characteristics	Original proposition	Behavioural proposition
Complexity	The greater the complexity, the greater the rate of change	The greater the number of attention centres, the greater the rate of change
Centralisation	The higher the centralisation, the lower the rate of change	The lower the feeling of participation, the lower the rate of change
Formalisation	The greater the formalisation, the lower the rate of change	The greater the reliance on factorisation, the greater the rate of change
Stratification	The greater the stratification, the lower the rate of change	The greater the stratification, the greater the persistence of subgoals, and the lower the rate of change
Volume	The higher the volume, the lower the rate of change	The greater the level of satisfaction, the lower the rate of change
Effectiveness	The greater the emphasis, the lower the rate of change	The greater the emphasis on local efficiency, the lower the rate of change
Job satisfaction	The higher the satisfaction, the greater the rate of change	The higher the degree of congruence between individual and organisational goals, the greater the rate of change

However, the extent to which group problem-solving is superior to individual problem-solving depends on the nature of the problem-solving process. On the one hand, unlike the group the individual span of attention is rather limited. On the other hand, while the human brain is capable of simultaneous processing of information, the effectiveness of group problem-solving depends on the ability of the group (or the organisation) to break

---

45. In the latter case we may be referring to departments or the organisation as a whole, depending on the level of aggregation.

down information in parts which can be easily transmitted and comprehended in a sequential fashion. This requires that the causal means-end relationships are relatively clear, and that the means-end chain is genuine in the sense that the means are independent of each other. The organisational division of labour and the existence of occupational specialities are means of achieving such factorisation (March & Simon, 1958, p.193) which allows the process of group problem-solving to proceed sequentially while there is simultaneous information processing at the level of the organisation. In consequence, the speed of problem-solving is positively related to factorisation, and the rate of change will be high to the extent that the elaboration of performance programs can be factorised.

Thus, contrary to the Hage & Aiken (1970) and Zaltman et al. (1973) proposition, the behavioural theory seems to indicate that the rate of change is positively correlated with *formalisation* due to what you may call a factorisation maxim, i.e. the notion that it is possible to increase the tolerance for interdependence by arranging a simple and manageable task environment, where communication becomes effective. However, to the extent that this implies a high degree of *centralisation*, the benefits from factorisation tend to be traded off, since the organisation member's decision to participate in organisational activities depends on the inducement-contribution mechanism. To the extent that the intraorganisational power differences fall outside the area of acceptance[46], the rate of change will be low. This is a rather "soft" version of the centralisation argument in section 2.2. Furthermore, if factorisation is related to a high degree of intraorganisational *stratification*, the set of goals become more complex and subdivided, and to the extent that the individual identification with the goals of the subunit is strong, factorisation leads to a differentiation of subgoals which, as argued previously, tend to persist detrimentally to change.

The *volume* argument in section 2.2 referred to success as a source of complacence, and as argued previously, the occasions for innovation do not occur when the level of satisfaction is not violated, although there may be a tendency towards the perception of a performance gap depending on the position of the organisation in the aspiration-achievement relation, cf. figure 12. In cases, where the level of satisfaction is violated and the inducement to engage in innovation is high, this inducement tends to be reinforced by *job satisfaction*, since the organisation member's decision to participate in an organisatio-

---

46. Which may also happen in the case of less visible power differences that prior to innovation actually *was* outside the area of acceptance: While they *ex ante* were less visible and thus somewhat outside the focus of attention of the organisation member, they *ex post* become visible and move into focus.

nal activity depends on the congruence between his goals and the goals of the organisation. To the extent that congruence is present, it will fall within the area of acceptance.[47]

Finally, the *effectiveness* argument may be restated like this: Since the organisation, from a behavioural point of view, is essentially an adaptive institution, change is more likely to be suggested by those parts of the organisation which deals with the extraorganisational environment and/or is deviced to absorb uncertainty. Conversely, change is less likely to be suggested by those parts of the organisation which are engaged in the most detailed performance programs, since they tend to become reinforced, as argued previously. To the extent that an emphasis on effectiveness relies on detailed performance programs, the probability of innovation proposals decreases, and the rate of change is low. This is a type of restatement, which refers more to the routine nature of organisational activities than the argument presented earlier in section 2.2.

---

47. Actually, the behavioural theory discusses a number of aspects of congruence, such as the relationship between job characteristics and the individual's self-characterisation, supervisory practice and employee independence, and private and organisational roles. However, the present study does not delve into the motivation aspects of the behavioural theory to a greater extent than already implied above.

Chapter 4

Contingent Decision Making and Problem Solving

4.1. Integration and differentiation

How can integration be facilitated without sacrificing the needed differentiation ?

This is the central question posed by Lawrence & Lorsch (1967, p.13), who were the first to use the term "contingency theory". Differentiation refers to the affinity between the extraorganisational environment and the intraorganisational configuration, where the organisation is segmented into subunits each with the ability to deal with a certain segment of the environment. Integration refers to the ability of the organisation to overcome the potential conflicts entailed in differentiation. Inspired by Woodward (1958) and Burns & Stalker (1961), Lawrence & Lorsch (1967) argue that there is a positive relationship between the variation of the environment and the differentiation of the organisation, and the more differentiated the organisation the greater the potentials for intraorganisational conflicts and thus the need for coordination.

Studying the emergent Scottish electronics industry and comparing it to eight British firms, which were larger and more committed to the development and manufacture of electronics because they had a longer historical record within the industry[48], Burns & Stalker (1961) identified two major, ideal-type systems of management practice, cf. figure 18. In stable task environments, the organisation tends to assume a mechanistic form which is characterised by hierarchial coordination based on functional specialisation and vertical interaction. In volatile task environments, the organisation tends to assume an organic form which is characterised by network coordination based on the continual redefinition of tasks and horisontal interaction. *bureaucracy*

While the mechanistic form resembles the ideal-type, Weberian bureaucracy of impersonal relationships within a highly predictable task environment, the organic form is the negation of bureaucracy in the sense that the organisation members must invest themselves as distinct human beings in a mutual interpretation of the task environment. Thus, while the mechanistic form relies on instructions and decisions as the focus of communication, the organic form relies on consultative information and advice.

---

48. In sum, twenty firms were studied. The firms differed considerably according to size, but also affiliation, i.e. the sample contained firms which were part of a larger concern. However, Burns & Stalker (1961) attempted to isolate the relevant functions into "an organized working community, concerned with designing, making, and selling products which incorporate electronic components" (ibid., p.77). *stable & volatile task environments*

Knowledge in the mechanistic organisation tends to become located at the top of the hierarchy, where decisions are made and diffused downwards as top-down instructions; contrary, knowledge may be dispersed across the organic organisation, and the location of knowledge tends to become an ad hoc centre for control, coordination and communication. Hence, participation of managers is greater in the organic than the mechanistic organisation, and so is the reliance on personal commitment *vis-á-vis* the emphasis on loyalty in the mechanistic setting.

Figure 18. *Organic and mechanistic forms of management practice*

Feature	Organic	Mechanistic
Overall design	Network structure of control, authority and communication	Hierarchial structure of control, authority and communication
Task environment	Contributions from special knowledge and experience of organisation members	Specialised differentiation according to how problems are broken down
Organisation roles	Individual tasks are defined by the total situation of the organisation	Individual tasks are determined by functional rationality
Communication focus	Information and advice	Instructions and decisions
Coordination	Individual tasks area adjusted through lateral interaction	Hierarchial planning and control through vertical interaction
Participation	Shedding of responsibility and decision on the basis of an insistence on commitment	Precise definition of roles according to function on the basis of an insistence on loyalty
Focus of knowledge	Knowledge located anywhere	Knowledge located at the top

Source: Burns & Stalker (1961, pp.119-22)

However, as argued by Burns & Stalker (1961, p.122), the two distinct systems of management practices "represent a polarity, not a dichotomy". A number of intermediate positions exist, and an organisation which is "oscillating between relative stability and relative change may also oscillate between the two forms" (ibid.). This process of oscillation represents a number of shifts between the reliance on programmed and nonprogrammed decision making. From a behavioural point of view, this may be paraphrased by saying that the relative importance of substantial and procedural coordination tends to shift, and the greater the stability of the extraorganisational environment, the greater the reliance on substantial coordination.

The comparison between the electronics firms and two firms within, respectively, the rayon and the switchgear industries (Burns & Stalker, 1961, pp.77-95) illustrates this point, cf. figure 19. In a stable environment characterised by a very small rate of technical change along familiar patterns (the rayon

— familier

case), the productive activities are undertaken according to stable programs, and decision making is based on "the framework of familiar expectations and beliefs" (ibid., p.83). Tasks are relatively clearly defined and deviced to maintain the conditions of production stable. In a slightly stable environment characterised by a moderate rate of technical change based on costumer requirements (the switchgear case), there appears to be a combination of mechanistic and organic forms. While the activities of production and design are fairly standardised, the coordination of management decision processes is based on frequent interaction and continual reinterpretation of expectations and opportunities. Furthermore, the work flow at the

Figure 19. *Intermediate organisations*

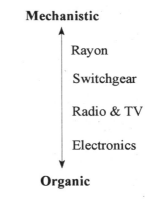

Mechanistic

Rayon

Switchgear

Radio & TV

Electronics

Organic

Source: Inspired by Morgan (1986, pp.52-53)

shop levels tends to be distributed in a non-routine fashion. In a slightly unstable environment characterised by rapid technical change with a predictable rate of novelty (radio & TV), the responsibilities of managers become less clearly defined, and influence on the decision process resides in knowledge and competence rather than hierarchial positions. Finally, in a rather unstable environment characterised by rapid technical change with an unpredictable rate of novelty (electronics)[49], the managerial positions are only specified to the smallest possible extent, and managerial roles are "continually

Compare to K-12 hierarchy

49. In most of their analysis, Burns & Stalker (1961) include the radio & TV business in the electronics industry. However, in their comparisons across industries, the electronics industry is divided into radio & TV and electronics "proper".

defined and redefined in connexion with specific tasks and as members of co-operative groups" (ibid., p.93). This extremely interactive mode of management coordination is highly sensitive to the efficiency of communication.

The implication of the Burns & Stalker (1961) inquiry is that the management practice of an organisation vary with the characteristics of the environment[50], and that each organisation form is distinct regarding the means of coordination and integration. Lawrence & Lorsch (1967) elaborate on this proposition and suggest that the differences in management practice are reflected in differences between organisations regarding the means by which conflict is resolved. Studying six companies in the plastics industry and comparing them to four companies in the container and packaged food industries[51], cf. figure 20, Lawrence & Lorsch (1967) suggested that the most succesful were those which achieved an appropriate degree of differentiation *vis-á-vis* the extraorganisational environment and a high degree of quality of intraorganisational integration. *Differentiation* is defined as "the difference in cognitive and emotional orientation among managers in different organizational departments" (ibid., p.11) and comprises four variables: The managers' orientation towards specific goals; the time orientation of managers, i.e. the time horizon of problem solving; the interpersonal orientation among managers, i.e. the way

Figure 20. *The Lawrence & Lorsch (1967) study*

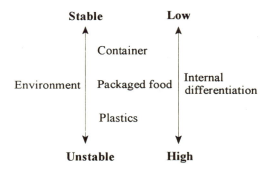

in which they relate to one another in problem solving; and the formality of the structure of the functional units. Differentiation reflects that certain parts of the organisation is designed to deal with certain parts of the task environment. For instance, we might expect that "research units, dealing with highly uncertain tasks, would be less formally structured than the production units", where "production managers, to do their jobs properly, would

---

50. Which is measured in terms of technical change and market growth.

51. The six plastic organisations were divided into high, medium and low performers, equally distributed. The container and packaged food organisations were pairs of a high and a low performer. Performance was measured in terms of change in profits and turnover, and the share of new products in current sales, during the past five years.

require more established routines and tighter controls" (ibid., p.31). Furthermore, we might expect that production managers are focussed on short-run production issues, while research managers are focussed on long-run problem solving.

Integration is defined in terms of the means employed in and the nature of conflict resolution, which are analysed with respect to six determinants, three of which relate to the behaviour of integrators, and three of which relate to the behaviour of functional specialists as well, cf. figure 21. Integrators are managers with the responsibility of linking together the activities of the functional departments. Integrators may reside in their own department designed for that purpose, or in the functional departments at the level appropriate for the interaction with managers in those departments that are to be linked.

Figure 21. *Determinants of conflict resolution in decision making behaviour*

Determinant	Definition
Integrators	
Intermediate position	The distance between the integrator's goal, time, and interpersonal orientation and those of the managers in the departments he is linking
Influence	The degree to which the integrator is percieved as influential by the managers in the departments he is linking
Reward system	The extent to which the integrator is rewarded on the basis of his ability to achieve integration between the departments he is linking
Integrators and functional specialists	
Total level of influence	The extent to which the department in question is percieved as influential by the relevant agents of that department
The center of influence	The extent to which the influence is concentrated at the managerial level where the knowledge to make decisions is available
Modes of conflict resolution	The extent to which causes of conflicts are - eliminated in open discussion: *Confronting* - compromised or neglected: *Smoothing-over* - eliminated by way of authority or coalition: *Forcing*

Source: Lawrence & Lorsch (1967, pp.73-78; 265-67)

Regarding the quality of integration, one should observe the following in relation to figure 21.

First, integration becomes effective to the extent that the integrator's orientation towards goals, time horizon, and interaction is similar or very close to the orientation of the managers of those departments which he is supposed to link. If this is the case, conflict resolution as the creation of a set of common beliefs is facilitated. The motivation of the integrator to achieve effective integration is positively stimulated if his results are evaluated on the basis of integration rather than on the basis of some other criteria, e.g. seniority or overall productivity. Furthermore, the quality of integration is positively related to the percieved influence of the integrator, i.e. the managers with which the integrator interacts are more likely to play an affirmative role in the process of integration if they consider the integrator to be able to promote the mutually reached solutions within the organisation.[52]

Second, to the extent that the participants (departments) in the integrative process feel that they are able to influence the decision making process, they will be inclined to support the outcome and prove less hostile to other participants (departments). Furthermore, their inclination to contribute to conflict resolution is stimulated if they feel that their knowledge about possible solutions is matched by the appropriate degree of influence. Finally, to the extent that the causes of conflicts are recognised through mutual acknowledgement, the resolution of conflict tends to become more effective.

4.2. Combining the logic of rational and natural system models

The seminal studies by Burns & Stalker (1961) and Lawrence & Lorsch (1967) involves several theoretical implications. *First*, they represent the antithesis to the classic point of view that one may find "the one best way to organize in all situation" (Lawrence & Lorsch, 1967, p.3). On the contrary, there is "no one best way to organize" (Galbraith, 1977, p.28). Organisations tend to survive if they match extraorganisational requirements, and the extraorganisational environments tend to differ across the economic system.

---

52. Lawrence & Lorsch (1967) emphasises the percieved influence, since they argue that the positive relationship between integration and the integrator's influence is not solely dependent on the actual influence of the integrator. If the linked managers *believe* him to be influential, they will take an active part in the process of integration, although this belief may not be objectively true. However, one could argue that if an integrator is judged as influential to an extent which make him able to link the activities of other organisation members with their approval, he does, in fact, become influential. On the other hand, there is, of course, a difference between the ability to persuade, for instance, colleagues at the horisontal level to do certain things and to persuade superiors at the vertical level to approve and support these actions.

Second, they indicate that as the extraorganisational environment becomes more volatile, the natural system model assumes predominance over rational system models, as indicated in chapter 1. To the extent that the extraorganisational environment confronts the organisation with uncertainty, the organisation must adapt in a way that constitutes it as a natural phenomenon in the sense proposed in figure 3.

However, *third*, this does not imply that the organisational configuration tends to adhere to the types of designs suggested by the natural system approach. On the contrary, the most viable organisations tend to exhibit various types of intraorganisational rationality according to the nature of the departmental task environments. This third, differentiated implication, which make it possible to combine rational and natural system models, is at the focus of analysis in the work by Thompson (1967) that appeared simultaneously with Lawrence & Lorsch (1967). According to Thompson (1967), the rational and natural system models reflect, respectively, a closed-system and an open-system strategy for the study of organisational behaviour, where the former tends to ignore "uncertainty to see rationality", while the latter tends to "ignore rational action in order to see spontaneous process" (ibid., p.10).[53] However, just as the closed-system and open-system strategies are present in organisation studies, they are present in the strategies of management for organisational design. In consequence, Thompson (1967) suggests a three levels model of the distribution of responsibility and control[54], cf. figure 22.

According to this model, the organisation tries to impose a condition near certainty at the technical level by removing as many of the contingencies creating uncertainty to the managerial and institutional levels. To the extent that the nature of the extraorganisational environment and the technical level differ across organisations, we should expect the nature of the managerial and institutional levels to differ as well, and relating his work to Simon (1957) and Cyert & March (1963), Thompson (1967) suggests that the adaptive processes of searching, learning and deciding vary with the differences in intraorganisational technology and the extraorganisational environment.

The definition of technology applied by Thompson (1967) is rather broad. Similar to the simultaneous work by Perrow (1967), Thompson (1967) regard technology as the defining characteristics of organisations and describes it in terms of principles of coordination, i.e. the way in which the interdependence of organisation members are de-

---

53. Thompson (1967, p.10) attributes this dichotomy to "the fact that our culture does not contain concepts for simultaneously thinking about rationality and indeterminateness".

54. Inspired by Talcott Parsons (1960), *Structure and Process in Modern Societies*, New York: The Free Press of Glencoe.

Figure 22. *The three levels model of Thompson (1967)*

Level of activity	Content of activity
Technical	Focussed on the effective performance of the technical function, i.e. the transformation of inputs into outputs
Managerial	Mediating between the technical function and those who use its outputs Procuring the resources necessary for the technical function Controlling the technical function Mediating between the technical and the institutional levels
Institutional	Relating to the extraorganisational environment by establishing organisational boundaries and legitimacy

Source: Thompson (1967, pp.10-12) and Scott (1992, p.99)

signed[55]. Intraorganisational interdependence may be pooled, sequential or reciprocal. *Pooled* interdependence refers to the situation where "each part renders a discrete contribution to the whole and each is supported by the whole" (ibid., p.54), as in the case of a multi-divisional firm with partially independent branches that share resources. *Sequential* interdependence refers to the situation where the output of one unit becomes the input of another, as in the case of related points at an assembly line or a supplier-user relationship:

---

55. Perrow (1967, p.195) defines technology as "the actions that an individual performs upon an object, with or without the aid of tools or mechanical devices, in order to make some change in that object. The object, or *raw material*, may be a living being, human or otherwise, a symbol or an inanimate object. People are raw materials in people-changing or people-processing organizations; symbols are materials in banks, advertising agencies and some research organizations; the interactions of people are raw materials to be manipulated by adminstrators in organizations; board of directors, committees and councils are usually involved with the changing or processing of symbols and human interactions, and so on" (quotation marks in the original text substituted by italics). Thompson (1967, pp.15-18) distinguishes between *long-linked* technology characterised by serial interdependence as in the case of the mass assembly line; *mediating* technology which links independent organisational clients or customers as in the case of insurance firms and post offices; and *intensive* technology where the selection, combination and order of application of techniques at the object depends on the feedback from that object, as in the case of a general hospital or the construction industry.

Here both make contributions to and are sustained by the whole organization, and so there is a pooled aspect to their interdependence. But, in addition, direct interdependence can be pinpointed between them, and the order of that interdependence can be specified.

(Thompson, 1967, p.54)

Finally, *reciprocal* interdependence refers to the situation where "the output of each become inputs for the others" (ibid., p.55), as in the case of the relationship between maintenance and operation units within an airline, where the transportation of passengers relies on the maintenance service but at the same time provides the input for maintenance through wear and tear of the aeroplane.

The three types of interdependence form a hierarchy in two respects. *First*, they form a hierarchy of increasing degrees of interdependencies in the sense that reciprocal interdependence involves sequential and pooled interdependence, while sequential interdependence involves pooled interdependence. *Second*, as depicted in figure 23, the three types of interdependence differ according to the difficulty of coordination. Now, the hierarchy is reversed according to increasing contingency. The aspect of contingency is most important in the case of reciprocal interdependence where each unit has to adjust its

Figure 23. *The Thompson (1967) model of interdependence and coordination*

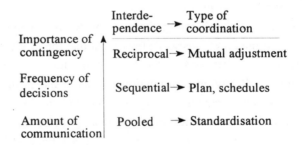

acitivities to the activities of the other; less important in the case of sequential interdependence where the adjustment depends on whether or not the output of a preceding activity meets the requirements of the proceeding activity; and least important in the case of pooled interdependence where the activities of one unit are independent of the other units as long as the mutual relationship is not jeopardised. Accordingly, the kind of coordination that seems appropriate vary with the importance of contingency and thus the frequency of decision making and the amount of communication necessary to undertake decision. In the case of *pooled* interdependence, where the importance of contingency is the least, coordination by a set of rules or routines becomes feasible, i.e. coordination by standardising the way in which decision making is undertaken, cf. section 3.2. In the case of *sequential* interdependence, where the importance of contingency is medium and the degree of stability and feasible routinisation of the situation is less than in the case of pooled interdependence, coordination by plan becomes feasible, i.e. "the establishment of schedules for the interdependent units by which their actions may then be governed" (ibid., p.56). Finally, in the case of *reciprocal* interdependence, where the importance of contingency requires that new information is transmitted among units during the process of action, coordination by mutual adjustment, i.e. feedback, seems appropriate.

In consequence, in order to minimise the costs and time consumption of coordination, the coordination of pooled interdependence may rely on pervasive rules applied to the widest possible category of units, while the coordination of sequential and reciprocal interdependence may rely on a clustering of positions into the smallest possible units. These clusters may then become coordinated by the standardisation strategy of pooled interdependence, which requires that a number of liason positions, i.e. the integrators of Lawrence & Lorsch (1967), are established.

The way in which this intraorganisational coordination is achieved reflects the complexity of the environment. Thompson (1967, pp.70-73) describes the extraorganisa-tional environment according to whether it is homogenous or heterogenous, stable or shifting[56], and argues that the organisation tries to deal with the task environment by, to the largest possible extent, isolating the technical core and the boundary-spanning units from each other and subjecting them, respectively, to the closed- and open-system strategies, as argued previously. To the extent that the task environment is homogenous and stable, or the variations in the task environment are predictable, the boundary-spanning units will rely on coordination through a standardised set of rules. If the task environment is heterogenous, the organisation will subdivide in order to create a set of homogenous task environments each to be dealt with by "rule-applying agencies" (ibid., p.72). If the variations of the task environment are large and somewhat unpredictable, there is less reliance on coordination by rules, and to the extent that the task environment becomes heterogenous, differentiation occurs in order to achieve adaptability through flexibility.

---

56. Thus comprising the views of Burns & Stalker (1961) and Lawrence & Lorsch (1967). However, Thompson (1967) does not refer to Burns & Stalker (1961), and Burns, Lawrence, Lorsch and Stalker do not enter his references.

However, in both cases *uncertainty* may occur, not only by *extraorganisational* but also by *intraorganisational* contingency to the extent that knowledge or beliefs, and preferences about cause-effect chains and solutions are uncertain in the way depicted in figures 15-16. In order to be able to make effective decisions, the organisation may apply a number of decision strategies depending on the nature of these beliefs and preferences, as summarised in figure 24.

Figure 24. *Strategies of decision making in relation to uncertainty*

Knowledge or beliefs about cause-effect chains	Preferences regarding possible outcomes	
	Certain	Uncertain
Certain	Computational strategy	Compromise strategy
Uncertain	Judgmental strategy	Inspirational strategy

Source: Thompson (1967, pp.134-35)

These strategies for decision may be explained in terms of figures 15-16. The *computational* strategy is similar to the programmed decision making described by the behavioural theory of the firm, where it is possible to assess whether a solution is good or poor. Thus, the computational strategy may be applied to the case of unacceptability, and of course to acceptability as well. The *compromise* strategy is applicable in the case of incompatibility and relates to the case of bland assessment, while the *judgmental* strategy is necessary in the reverse case which relates to mixed assessment. The degree of certainty about preferences in these two cases may be explained by the dispersion of influence. If the influence on organisational decision making is widely dispersed among the organisation members, preferences about outcome become less clear, and a compromise strategy must be applied. If the influence tends to be concentrated, a judgmental strategy occurs. Thompson (1967) makes a number of propositions regarding the dispersion of influence and the power of the dominant coalition, such as:

> The more numerous the areas in which the organization must rely on the judgmental decision strategy, the larger the dominant coalition. [...] Where technology is incomplete or the task environment heterogenous, the judgmental decision strategy is required and

control vested in a dominant coalition. [...] When power is widely distributed, an *inner circle* emerges to conduct coalition business. [...] The organization with dispersed bases of power is immobilized unless there exists an effective inner circle. (...) When power is widely dispersed, compromise issues can be ratified but cannot be decided by the dominant coalition in toto.

(Thompson, 1967, pp. 136; 143; 140; 141)

Finally, the *inspirational* strategy must be applied in the case where outcomes as well as preferences are uncertain, i.e. an experimental strategy of decision making which relies more on hints and intuition than on the certainty of knowledge.

In sum, uncertainty occurs from intra- and extraorganisational contingency in relation to how intraorganisational local rationalities, environmental demands, and the rationality of the organisation and its environment are reconciled. Characterising this conclusion in terms of the flexibility-stability approach, one may argue the following:

On the one hand, sealing off the technical core from contingencies in order to secure smooth operation calls for a computational decision strategy which requires that preferable cause-effect relationships and outcomes have been formulated at the level of tactical and strategic planning. On the other hand, the organisation as a whole must be able to flexibly adapt to environmental changes, and in consequence the goals formulated at the managerial level are characterised by diversity in order to deal with a large number of contingencies. Thus, the organisation is faced with a conflict between computational and non-computational rationality.

Thompson (1967, pp.148-50) refers to this immanent conflict as the *paradox of administration*:

In the *short run*, administration seeks the reduction or elimination of uncertainty in order to score well on assessments of technical rationality. In the *long run*, however, we would expect administration to strive for flexibility through freedom of commitment - i.e., slack - for the larger the fund of uncommitted capacities, the greater the organization's assurance of self-control in an uncertain future. (...) ...the time dimension of concern is inversely related to the level in the organization's administrative hierarchy. Thus at the upper reaches, or institutional level, we would expect the short run to be relatively insignificant and the longer run to be of central concern. Here we would expect the focus to be on increasing or maintaining flexibility and command uncommitted or easily recommitable resources. At the other extreme, the technical core, we would expect concern for certainty in the short run to drive out consideration of the longer run. The central part of the administrative hierarchy, the managerial layer, would thereby become the "translator," securing from the institutional level sufficient commitments to permit technical achievement, yet securing from the technical core sufficient capacity and slack to permit administrative decision and, if necessary, recommitment of resources.

(Thompson, 1967, p.150)

As we may recall from section 2.3, these considerations leads Thompson (1967, pp.151-52) to suggest *opportunistic surveillance* as a style of search behaviour alternative to the problemistic search described in section 3.2. Opportunistic surveillance is proactive in that it "does not wait to be activated by a problem and which does not therefore stop when a problem solution has been found" (ibid., p.151). However, opportunistic surveillance, which Thompson (1967) expect to find at the institutional level, occurs only on rare occasions or in special circumstances, for several reasons (Thompson, 1967, pp.152-54): *First*, administration may be percieved as "officeholding", i.e. the adminstrative members may see themselves as occupying a position rather than playing an active role in the organisation. *Second*, there may exist a bias towards certainty reflected in the predominance of short-term considerations over long-term considerations, e.g. associated with a lack of distinction between managerial and institutional matters, and an intolerance for ambiguity, a phenomenon discussed in chapter 6. *Third*, the nature of the coalition may be such that no one is empowered to give direction to the organisation, because the influence is too dispersed and no inner circle emerges. *Four*, there may simply exist a lack of knowledge which confines the organisation to incrementality or change through long-term trial-and-error.

Figure 25. *A routine/non-routine perspective on search and problem solving*

Problem solving	Search	
	Problemistic	Opportunistic
Reproductive	Routine	Quasi non-routine
Productive	Non-routine	Non-routine

Source: Gjerding (1992, p.108), table 5.1

The rareness of opportunistic surveillance may also be explained in terms of routinisation of organisational activities, cf. figure 25, which, however, implies that opportunistic surveillance occur more frequently than suggested by Thompson (1967). Figure 25 reflects the previous argument that reproductive and productive problemistic search are, respectively, associated with programmed and nonprogrammed activity, and suggests that the search behaviour entailed in opportunistic surveillance may be described in the same vertical terms. Reproductive opportunistic surveillance describes the situation in which the organisation member tries to predict future problems and alternative solutions by applying knowledge that are already present in the organisation or familiar to him. The attempt to anticipate problems implies that we are dealing with non-routine behaviour because the creation of new activity programs is involved. However, these programs are percieved by the aid of known heuristics and thus represent incremental changes. In conclusion, the combination of opportunistic search and reproductive problem-solving reflects a quasi non-routine activity. A movement downwards to the south-east quadrant in figure 25 occurs when the direction of opportunistic search implies the application of new knowledge unfamiliar to the organisation member.

To the extent that opportunistic surveillance becomes a programmed activity based on reproductive problem solving, it may satisfy the bias towards certainty. Furthermore, it may lead to a stream of innovation proposals which may help to overcome the problem of the power stalemate described by Thompson (1967): Reproductive opportunistic surveillance will contribute to the accumulation of knowledge, and as argued by e.g. Burns & Stalker (1961) the possession of knowledge about improvements of the organisational activities is, in itself, a power resource. Thus, reproductive opportunistic surveillance may aid the formation of an inner circle empowered to give direction to the organisation. On the other hand, to the extent that administration is percieved as "office-holding", reproductive opportunistic surveillance will be impeded unless the programmed search is based on some sort of incentive structure and performance indicators, e.g. as suggested by March & Simon (1958).

4.3. Implications to the flexibility-stability dilemma

Summing up on the review of the three seminal studies of Burns & Stalker (1961), Lawrence & Lorsch (1967) and Thompson (1967), the contingency propositions with respect to the flexibility-stability dilemma may be stated as follows:

1. The performance of organisations depends on their ability to achieve an appropriate match between the organisational configuration and the extraorganisational task environment. The degree of environmental variation must be met with an appropriate degree of organisational differentiation.

2. Organisational differentiation creates a diversity of intraorganisational perspectives which may lead to conflict, both when business is conducted as usual and when the organisational activities are changing. Organisational differentiation requires an appropriate quality of organisational integration based on adequate conflict resolution. Conflict resolution becomes effective, when the distance between organisation members' perceptions is small and integration is performed at the appropriate levels of influence.

3. The diversity of intraorganisational perspectives implies that different subunits may be subjected to different types of administrative rationality, as indicated by the theoretical division between rational and natural system models. The range of subunit rationality is determined by technology defined as the intraorganisational division of labour and type of coordination. The difficulty of coordination is positively related to the importance of contingency and thus to the frequency of, and the amount of communication necessary to, decision.

4. The way in which decisions are undertaken depends on the degree of certainty about cause-effect chains and organisational preferences. Furthermore, decision making is embedded in a flexibility-stability paradox of administration which implies that problem solving should be connected to nonprogrammed search at the managerial and institutional levels, while programmed search may be feasible at the technical level. However, search tend to be problemistic in the sense of Cyert & March (1963).

5. Uncertainty appears as a gap between the knowledge of the organisation members and the knowledge necessary to cope with contingencies. In order to deal with this information gap, organisational rationality becomes, essentially, bounded in the sense of Simon (1957) and March & Simon (1958). The organisation members find themselves within a framework of delimited responsibility, control and resources, where the degree of uncertainty is reduced or redefined to a manageable level where only those contingencies relevant for the individual task environment are included. However, there exist various types of bounded rationality within the same organisation according to the types of technology which the organisation applies, as implied by statement 3.

The notion of uncertainty as a gap between existing and required knowledge, or information, has been employed by Galbraith (1977) in his work on organisation design.[57] The basic idea is based on the Simonian notion of bounded rationality: As uncertainty increases, so does the number of exceptions to the existing organisational repertoire and thus the number of decisions which are referred upwards in the organisational hierarchy. This process confronts the upper levels of management with information overload in the sense that the amount of information which they have to process exceed their information

---

57. "Uncertainty is the difference between the amount of information required to perform the task and the amount of information already possessed by the organization" (Galbraith, 1977, pp.36-37). Galbraith (1977) focusses on the aspect of information and systems of information flow as a reminiscence of his early occupation with computer technology in the mid-sixties. However, he defines himself as intellectually endebted to Thompson: "He is the one responsible for my focusing on organizations rather than computers. I was extremely fortunate to be able to take his doctoral seminar as he was writing *Organizations in Action*. The class discussions of each new chapter gave me an appreciation of the man as well as the material" (Galbraith, 1977, p.xi).

processing and computational capacities. Essentially, the organisation is faced with two options in order to overcome the problem of information overload: It may reduce the need for information processing by "reducing the level of goal performance, the division of labor, or the diversity of goals" (ibid., p.81); or it may increase its ability to process information. Each option is connected with a number of organisation design strategies (ibid., p.49): The need for information processing may be reduced through environmental management, the creation of slack resources, or the creation of self-contained tasks; the capacity to process information may be increased by investments in vertical information systems or the creation of lateral relations. Whichever strategy chosen, the objective is to reduce the number of exceptions which are referred upwards in the hierarchy.

Environmental management refers to the attempts to control the external environment, either by influencing the decision processes in extraorganisational units, or by integrating parts of the environment which are especially critical to the functioning of the organisation. An example of the former is the attempt of the organisation to influence the decisions of consumers by public relations, or to take political actions in order to influence the way in which markets are regulated (Galbraith, 1977; Schein, 1988). An example of the latter is the increasing propensity to place R & D activities within the boundaries of the organisation (cf. Teece, 1988), or to vertically integrate unreliable sources of inputs (Schein, 1988).

Creation of slack resources refers to the reduction of exceptions, which can be attained if the required levels of performance are reduced. For instance, a sales department may present the costumers with longer delivery times, if the production department frequently experiences difficulties with achieving promised delivery times; or, the problem may be solved by hiring additional workers and buy additional psychical capital, which are alternately under- and over-utilised (Galbraith, 1977). In the field of economics, it is generally assumed that firms respond to problems of capacity by utilising its input resources more efficiently. However, Galbraith (1977) argues that this strategy may increase the interdependence between the constituent parts of the organisation. Increasing interdependence may, in itself, create uncertainty because of bounded rationality, i.e. the limits on information processing and computational capacity (Simon, 1957). The creation of slack resources reduces the intraorganisational complexity and thus problems to a level which can be managed, and may therefore be "less costly than other alternatives" (Galbraith, 1977, p.81).

Creation of self-contained tasks refers to the reduction of diversity and division of labour by collecting resources which are able to deal with all the aspects of a limited task environment. Creating self-contained tasks "shifts the basis of the authority structure from one based on input, resource, skill, or occupational categories to one based on output or geographical categories" (Galbraith, 1977, p.51). The self-containment strategy implies the creation of subunits, each with the sufficient number of specialties required for the task, and results in the reduction of conflicts often associated with the utilisation of the same type of resources across task environments. For instance, Galbraith (1977) argue that the establishment of three self-contained units within a professional organisation which each serves it own client group reduces conflicts about priorities as compared to a configuration where a number of specialties provided service *across* the three client groups. Furthermore, the organisation increases its information processing capacity, since "the point of decision is moved closer to the source of information. Exceptions have to travel through fewer levels before reaching a shared superior" (ibid., p.85).

Investment in vertical information systems refers to the creation of information gathering and processing procedures which collect information at the points, where it originates, and disseminate information to these parts of the hierarchy where the decisions are to be made. The creation of vertical information systems, which Galbraith (1977, pp.109) relates to information technology as "an organizing mode", rests on the basic argument that in many cases it may be time- and cost-saving to device new plans instead of adjusting existing plans in the face of new information and exceptions, since goals and plans tend to become obsolete. Investments in vertical information systems requires that certain parts of the organisation are devoted to the collection, analysis, and dissemination of information. However, as pointed out by Butler (1991, p.160), the increase of the information processing capacity may be traded off by increasing managerial costs.

Finally, the *creation of lateral relations* refers to the employment of decision processes across the lines of authority in order to delegate discretion to the lower levels of the hierarchy and move decision making closer to the points of origin of information. Discretion at the lower levels of the hierarchy requires lateral relations to the extent that organisational resources are shared, in order to coordinate the use and control the costs of these resources. Galbraith (1977, pp.112-13) defines seven types of lateral relations:

1. Utilize *direct contact* between managers who share a problem.
2. Establish *liason roles* to link two departments which have substantial contact.
3. Create temporary groups called *task forces* to solve problems affecting several departments.
4. Employ groups or *teams* on a permanent basis for constantly recuring interdepartmental problems.
5. Create a new role, an *integrating role*, when leadership of lateral processes becomes a problem.
6. Shift from an integrating role to a *linking-managerial role* when faced with substantial differentiation.
7. Establish dual authority relations at critical points to create the *matrix design*.

These seven types of lateral relations vary according to the level of uncertainty and the complexity of interaction involved in the organisational activities, and their numbering reflects an increasing number of contingencies and a decreasing degree of predictability of tasks. Thus, the lateral relations represents various solutions to the problem of tolerance for interdependence described in section 3.2 (see figure 11) and gives, in my eyes, a more dynamic account of the tolerance for interdependence than March & Simon (1958).

Referring to Lawrence & Lorsch (1967), Galbraith (1977, p.151) argues that differentiated organisations face the problem of obtaining "overall task integration among departments *without reducing the differences that lead to effective subtask performance*", i.e. the problem of facilitating integration without sacrificing the needed differentiation, as stated by the opening sentence of the present chapter. As described in section 4.1, the use of integrators is a favourable solution to this problem. However, the integrating role may encounter a number of limitations which according to Galbraith (1977, pp.158-59) are related to "the limitations of expert power": First, the efficiency of the integrator role decreases as task uncertainty and thus the requirements of information processing increase. In this case, the efficiency of the integrator role is undermined by an increasing information gap, and we may describe this process as a gradual descent down the decision ladder depicted in figure 16, where the organisation eventually may find itself unable to judge the probability of a positively *and* negatively valued state of affairs.[58] Second, the organisation may find itself, from time to time, in situations where it is impossible to reach agreement on goals and thus give direction to the organisation. This is similar to the case of dispersed power described in the previous section.[59]

In consequence, the organisation may resort to managerial linking roles, as an elaboration of the integrating roles described by Lawrence & Lorsch (1967). Managerial linking roles differ from integrating roles in that they are established as normal managerial functions and devoted with approval power and the authority to enter the decision process at the initial stage of planning. Managerial linking roles have to be supported by vertical information systems if they are to be able to exert discretionary power, and the degree of discretionary power may be increased by allocating budget control to the managerial linking role. In sum, these managers

---

58. However, Galbraith (1977) neither uses this metaphor, nor refers to March & Simon (1958).

59. As the reader may recall, Thompson (1967) describes this case as one of the obstacles to opportunistic surveillance. The present study suggests that this obstacle may be overcome through reproductive opportunistic surveillance.

become planners, decision makers, and resource allocators. However, none of the resources is their responsibility. They have only information, knowledge, approvals, and money with which to influence the activities in their area of responsibility.

(Galbraith, 1977, p.161)

To the extent that the organisational activities are extremely differentiated because they rely on highly specialised resources in an unpredictable task environment[60], the power of integrators have to be increased. This is the case of a matrix organisation, where the integration takes place along two dimensions: Input and output.[61] The matrix organisation represents "a dual reporting relationship" (Galbraith, 1977, p.162) which stimulates search, and thus learning:

> In aerospace, project managers encourage performance within budget, on schedule, and within contract specifications. Laboratory managers encourage full utilization of resources, long-run resource development, and highly sophisticated technical performance. For some organizations these goals are of equal importance in general, but they vary in importance in specific instances. Each circumstance, which cannot be predicted in advance, needs to be resolved on its own merits. Rather than refer each circumstance to a general manager, the matrix design institutionalizes an adversary system. The resultant goal conflicts causes search behavior to discover current information and to create alternatives to resolve the conflict.

(Galbraith, 1977, p.163)

The matrix organisation highlights the importance of the flexibility-stability dilemma. Butler (1991, pp.167-69) argues that the risk of dysfunctions is high in matrix structures, since the "overriding problem of the matrix organization is to manage the dual logics of the technology and the environment" (ibid., p.167). The organisation members are confronted with different requirements from input managers and output managers: The former stresses the ability of the organisation member to stay ahead in his field of work, while the latter stresses the conformity to schedules and standards. In terms of figure 9, the organisation member is, simultaneously, confronted with a high degree of complexity *and* formalisation. Furthermore, the organisation member may find it difficult to locate the direction of organisational activities, since, as argued by Butler (1991), the risks of politicking is large because the decision making coalition tends to be large. Finally, conflicts are bound to occur because of the diversity of goals and perceptions described by Galbraith (1977).

---

60. Like in the case of aerospace, which Galbraith (1977) uses as an example.

61. Or processes and resources, and product, as Butler (1991, p.167) puts it.

Expressed in terms of figures 24-25, the way in which these conflicts are resolved within the matrix organisation is likely to combine two different types of decision making strategies. Take the normal case of a matrix organisation, i.e. an R&D intensive organisation. On the one hand, to the extent that R&D is associated with non-routine activities, search tends to be located at the productive opportunistic surveillance corner of figure 25. On the other hand, conflict resolution is not undertaken before conflicts actually occur. The conflict will relate to the diversity of goals and perceptions which are moulded by the productive problem solving that has taken place in the constituent parts of the organisation prior to the time of conflict. Thus, conflict resolution tends to involve problemistic search aimed at an assessment of the priority of existing goals in the light of the circumstances which have been created *ex ante* by productive problem solving. In consequence, the problem solving involved in conflict resolution tends to be located at the reproductive problemistic corner of figure 25, since the circumstances to which the conflict resolution relates are *ex post* phenomena. This means that while the search processes, which characterise the normal activities within the constituent parts of the matrix organisation, require a judgmental or inspirational decision making strategy, cf. figure 24, the search processes involved in the resolution of conflicts between the constituent parts require a compromise strategy. In other words, while the dimension of differentiation within a matrix organisation is likely to involve judgmental and inspirational strategies of decision making, the dimension of integration is likely to involve a compromise strategy.

The matrix organisation represents, of course, an extreme case of establishing lateral relations. However, the argument that the matrix organisation must rely on a combination of decision making strategies may be extended to cover the range of lateral relations. In each of the seven cases of lateral relations described above, the central problem of coordination relates to the management of a diversity of goals and perceptions, i.e. rationalities, which are brought together in a common task environment. Thus, the usage of lateral relations differ from the usage of slack resources and self-contained tasks in that the problem world of the organisation members are enlarged. Or, to put in the behavioural vocabulary, in the case of lateral relations the number of attention centres which interact is increased, while it is decreased in the case of slack resources and self-contained tasks. The greater the number of attention centres which are required to interact, the greater the risk of dysfunctional effects on organisational performance, and the smaller the tolerance for interdependence, unless the quality of integration is concomitantly increased. In consequence, the increase of the information processing capacity of the organisation is a double-edged option which requires that a suitable combination of decision making strategies is brought to balance at the knife-edge between dysfunctional and synergetic effects.

Although the description of the contingency theory in the present chapter has tried to stress the dynamics of organisational action in the sense that organisational action is moulded by a diversity of goals, perceptions and tasks, which create contradictory forces within the organisation, the dynamics presented by contingency theorists normally imply symmetrical relationships. The assumption of symmetrical relationships is present in, for instance, figures 20 and 23, and characterises most of the Galbrathian arguments. Schoonhoven (1981) has challenged the contingency assumption of symmetrical relationships by way of a model which tests three hypotheses entailed in Galbraith (1973), i.e. the hypotheses that effectiveness is positively associated with destandardisation, decentralisation and professionalisation the greater the degree of task uncertainty.[62] These propositions are symmetrical in the sense that uncertainty, organisational performance, and design variables move in the same direction. However, based on data from 8,593 patients who were treated by surgery in the operation room suites (OR) of 17 American acute-care hospitals, Schoonhoven (1981) questions the symmetrical nature of the contingency propositions. Instead, she suggests that the causal nature of each of the three relationships implied by the hypotheses may be reversed below a certain point of task uncertainty, i.e. the organisational performance becomes negatively associated with the design variables, and her data seemed to verify this proposition. There are several reasons for this verification (Schoonhoven, 1981, pp.370-71): Standardisation and centralisation increases control over outcome quality in low uncertainty task environments, since the hierarchy in these circumstances is able to satisfy the information processing requirements. Professionalisation creates problems associated with selective perception and the creation of variation: Since nurses are trained to judge on and manage uncertainty they tend to focus on uncertainty, and this may lead to misconceptions of patient characteristics and thus inappropriate choices; and since professionals qua their training may prefer variety, nurses may vary "their approach to the constant set of cases, with inconsistent results that reduce effectiveness in the long run" (ibid., p.371).

---

62. Destandardisation refers to the use of procedural rather than substantial problem solving, i.e. the less extensive use of performance programs; decentralisation refers to the location of discretion at lower levels of the hierachy; and professionalisation refers to the increase of skills in the work force, primarily by increasing reliance on craft or professional training (Galbraith, 1973). Effectiveness was measured by Schoonhoven (1981, p.360) in terms of "severe morbidity: a risk-adjusted postsurgical death and complication rate averaged for all patients undergoing surgery in the operation room suite of a given hospital".

This elaboration of the Galbraithian contingency propositions, which suggests that uncertainty relates to the design variables in two dimensions, is quite interesting since it seems to indicate that the search behaviour of highly skilled labour is inappropriate in low uncertainty task environments. For instance, while the OR workflow in cases of low uncertainty may be characterised by sequential interdependence, the problem solving behaviour of the professionals is more associated with the type of behaviour necessary in cases of pooled interdependence. Furthermore, although a low uncertainty OR workflow tends to make a computational decision making strategy more feasible, the professionals tend to adhere to a judgmental strategy. Finally, while reproductive problemistic search may be more appropriate, the professionals tend to employ productive problemistic search, or even opportunistic surveillance. As a consequence, diversity increases in conditions which are prone for uniformity.

Chapter 5
Two Cases of Organisational Innovation

which theory of, action applies.

5.1. An action research project

The behavioural and contingency theories represent different perspectives on organisational action, although they both belong to the class of rational system models. They differ
according to the degree to which the organisation is percieved as an open system. The
behavioural theory incorporates the notion of *external schocks* to an adaptive system, but
these schocks are primarily analysed as stimuli to adaptation or innovation. Contrary, the
contingency theory focusses on the *interaction* between the organisation and the
extraorganisational environment and argues that the organisation, in some cases, may be
able to influence the environment. Furthermore, the two bodies of theory hold different
perspectives on human nature and thus the nature of organisational behaviour. This may
be illustrated by the discussion of Morgan & Smircich (1980) on the ontological
assumptions within social science. They argue that differences in research methods may
be attributed to differences in the core ontological assumptions and assumptions about the
human nature which are depicted in figure 26. The notion of reality as a concrete *structure*
implies that the social world can be analysed as "composed of a network of determinate
relationships between constituent parts", where the human beings are

rational

> a product of the external forces in the environment to which they are exposed. Stimuli in
> their environment condition them to behave and respond to events in predictable and
> determinate ways. A network of causal relationships links all important aspects of behavior
> to context. Though human perception may influence this process to some degree, people
> always respond to situations in a lawful (i.e. rule-governed) manner.
>
> (Morgan & Smircich, 1980, p.495)

The notion of reality as a concrete *process* implies that the social world is evolving in a
way which expresses itself in contingent relationships: The "situation is fluid and creates
opportunities for those with appropriate abilities to mould and exploit relationships in
accordance with their interests"; the environments is interpreted by human beings who
engage in " a struggle between various influences"; and relationships "between individuals
and environment expresses a pattern of activity necessary for survival and well-being of
the individual" (ibid.). Finally, the notion of reality as a *contextual* field of *information*
implies that the social world is continously changing based on the transmission of

information, where human beings "are engaged in a continual process of interaction and exchange with their context", primarily based on learning through feedback (ibid.).

Figure 26. *Basic assumptions in some fields of social science*

Approach decreasingly objectivist, increasingly subjectivist, from left to right			
Core ontological assumptions	Reality as a concrete structure	Reality as a concrete process	Reality as a contextual field of information
Assumptions about human nature	Man as a responder	Man as an adaptor	Man as an information processor
Basic epistemo-logical stance	To construct positive science	To study systems, process, change	To map contexts
Some favoured metaphors	Machine	Organism	Cybernetic

Source: Morgan & Smircich (1980, p.492), table 1

The behavioural and contingency theories cannot be grouped unambigously into one of these categories.[63] The main emphasis of behavioural theory lies on the organisation member as a responder within the framework which determines the focus of attention. As argued by Morgan (1986), the work of March & Simon (1958) is primarily devoted to investigate the dysfunctional aspects of bureaucracy, but as classified by Peffer (1982), the behavioural theory is also a theory of intraorganisational political processes, where the dominant coalition struggles to preserve and increase its influence on the task environ-ment. Thus, reality enters the behavioural scheme as a concrete process as much as a concrete structure. The contingency theory clearly accepts the image of human beings as adaptive agents who exist in an interactive relationship with the environment, be it intra- or extraorganisational, and the same applies to the aggregate collective of organisation

bureaucracy + power

---

63. Morgan & Smircich (1980) are interested in the subjectivist-objectivist debate within the social science and employ a number of reality descriptions other than those in figure 26: Reality as a symbolic discourse, which is a basic assumption of social action theory; reality as a social construction, as in ethnomethodology; and reality as a projection of human imagination, as in phenomenology. These approaches have been omitted from figure 26, because they do not apply to the theories employed in the present study.

members, i.e. the organisation. However, the main emphasis, especially in the work by Galbraith, is on the processing and transmission of information in order to cope with uncertainty, accomodate to the changing environment, and to some extent infuse changes to the environment. Thus, reality enters the contingency scheme as a contextual field of information as much as a concrete process. In sum, the behavioural theory may be described as a structure/process approach and the contingency theory as an information-context/process approach. *Contingency theory*

The process dimension represents, in my eyes, the intersection between behavioural and contingency theorising. This point of view may be illustrated by the metaphoric approach of Morgan (1986), who describes the contingency theory by the organismic metaphor, the behavioural theory by the machine metaphor, and both theories by the brain metaphor. The machine metaphor refers to the conception of organisations as machine-like designed instruments for achieving certain ends. The organismic metaphor refers to the conception of organisations as living systems, often inspired by biological analogy. The brain metaphor refers to the conception of organisations as information-processing systems: *Gareth Morgan*

> Organizations are information systems. They are communications systems. And they are decision making systems. In mechanistic organizations these systems are highly routinized. And in matrix and organic organizations they are more ad hoc and free flowing. We can thus go a long way toward understanding organizations, and the variety of organizational forms in practice, by focusing on their information-processing characteristics.
>
> (Morgan, 1986, p.81)

In sum, behavioural and contingency theories may be described as part of the same body of theorising, since they both reflect rational system models and focus on organisational processes of information processing and decision making.

Retrospectively, a combined behavioural-contingency approach guided the way in which the two cases of organisational innovation reported in sections 5.2-5.3 was undertaken. Consequently, these cases may illustrate some instances of the flexibility-stability dilemma, and the remaining part of chapter 5 is devoted to this issue.

The word *retrospectively* is used deliberately:

The two cases of organisational innovation was part of a project on organisational change and human resource development induced by a vision of the practical implications of combining organisation design and psychological therapy. This vision developed out of a cooperation between two consultants, an organisation designer[63] and a psychological

---

63. Otto Bredsten, joint owner of the Danish consultancy Lisberg Management.

Investment in HR for process

therapist[64], who were inspired by the conclusions of Gjerding & Lundvall (1992) that contemporary process innovation requires the investment in human resource development, cf. chapter 8. Together with two medium-sized Danish manufacturing firms, these consultants initiated and took part in a project on the initiation and implementation of organisational change and human resource development as part of a change in the procedures of strategic planning within the two firms. The aim of the project, which was supported by the Danish Ministry of Education, was to secure the establishment of a new set of organisational roles based on the development of the social and technical competencies of the organisation members involved. The empirical evidence reported in sections 5.2-5.3 reflects a combination between case study and action research, since the gathering of evidence was an integral part of my involvement in the change process within the two firms. The evidence was not gathered with the purpose to illuminate some aspects of the flexibility-stability dilemma and the behavioural-contingency intersection, but reflects the occurrence of and attendance to the practical problems of the change process, which appeared during that process, and the gradual evolution of objectives posed by the managers and subordinates who participated in the project. _Retrospectively_, however, the evidence highlights a number of points made in chapters 2-4. This result may, of course, reflect the conceptual bias of those in charge of the project.[65]

The project consisted of four stages[66].

Stage 1, _analysis_, was devoted to the diagnosis of the organisational and human barriers to organisational and strategic change within the two firms. The main objective was to detect the hidden human resources of the firms and identify the needs for training and developing social competencies necessitated by the change process.

Stage 2, _implementation_, tried to implement the initiated changes by providing the necessary technical and social qualifications in a way that would _unfreeze_ the hidden human resources, resolve organisational conflicts, and create a common perspective and purpose of action among the organisation members involved. Due to feedback between analysis and experiences gained through implementation, stages 1 and 2 became intertwined.

---

64. Brita Lauridsen, who owns a one-woman consultancy.

65. The project was organised by Lisberg Management, and undertaken as a joint venture between the two firms, Otto Bredsten and Brita Lauridsen. Allan Næs Gjerding acted as external adviser and was in charge of the documentation and evaluation of the project. For reasons of anonymity, the two firms are referred to as A and B.

66. The stages have been documented to the Ministry of Education in subsequent mimeos by Allan Næs Gjerding.

Stage 3, *promotion*, was devoted to the generalisation, theoretically and practically, of the results, which is primarily done by Gjerding & Lauridsen (1995).

Finally, stage 4, *evaluation*, comprises the evaluation of the project undertaken at subsequent meetings with most of the organisation members involved in the project. In essence, only firm A participated in stage 4, and for this reason the empirical evidence obtained in firm A is the primary concern of this and the following section, while the experiences from firm B serve as comparative evidence.

At the stage of initiation, the two firms found themselves in different circumstances. The management of firm A had the intention to reorganise its sales department in a way which would create an entirely new type of procedural and substantive planning. The firm had no prior experience in large-scale reorganisation, and the organisation members involved found themselves in a situation with very few points of reference. Furthermore, the organisational changes were percieved as necessary in order to support a change in the way in which strategic planning was undertaken, but the firm had only a limited tradition for strategic planning as such. The organisation members involved in the change process in firm B were more familiar to strategic planning, since firm B had a tradition in this field, and for accommodating organisational changes as well. The change of the process of strategic planning in firm B did represent *something* new, but in the line of activities that the organisation members were used to undertake.

The difference between the organisational change in firms A and B might be associated with the difference between radical and incremental innovation in the sense that the organisational change in firm A was characterised by a larger degree of task uncertainty and obsolescence of routines, while the organisational change in firm B was characterised by the addition of new routines to the existing repertoire of strategic planning. The following diagnoses will make this description valid.

5.2. Problem: Resistance to organisational change

In firm A, a producer of large-scale heat exchangers for industrial purposes, various organisation members in the sales department had for some time contemplated on the benefits and deficits entailed in focusing on products rather than geographical markets. The sales department was grouped according to functions, and managerial responsibilities were allocated on the basis of geographical markets. However, the salesmen felt that they would appreciate the technical capabilities of the firm more, if they were able to concentrate on the promotion of product lines rather than the promotion of the product mix as a whole at various geographical markets. Some frustration was felt among the salesmen, and after several years of discussion, the human resource director initiated a questionnaire where the sales personnel, i.e. secretaries, technicians, and salesmen,

reorganization change

participated anonymously. The result showed a majority support for a reorganisation of the sales department according to product lines, and the human resource director deviced a plan which was presented to and supported by the board of directors.

The main objective of the plan was to delegate the custom-oriented activities to the national and global network of agents, i.e. the activities related to the dimensioning and calculation of costs and profits of deliveries, in order to release intraorganisational resources to the development of markets and product lines within the sales department. The plan divided the product mix into four main product lines according to the segments of industrial customers, i.e. manufacturing, food, energy, and shipping. Related to each segment, a sales department subunit would be established and include the tasks of salesmen, technicians, and secretaries, who were supposed to work together in a team-like fashion according to the set of tasks of the subunit rather than the set of traditional functional tasks.[67] Consequently, four subunit managers were appointed and they reallocated among them the sales department personnel according to subunit functions.

However, for a number of reasons, the formation of subunits became more difficult than anticipated, because the formal arrangements were undertaken at a greater speed than the necessary learning processes. The subunit managers found themselves with a staff that was not accustomed to work together as a team. Furthermore, they were supposed to master one worldwide product line rather than a number of geographical markets which involved several product lines, and this was an entirely new situation. The single subunit staff, on the other hand, faced a new manager and awaited his first attempts to coordinate the team. Consequently, the new subunits lacked direction, and the activities of the subunits tended to stall before take-off.

During the stage of analysis, it became clear that the lack of direction created a state of paralysation of the organisation members, managers as well as staffs. A number of talks between the consultants and the organisation members, individually and together in the subunits, revealed that most of the organisation members tended to focus on functional tasks according to their background of training and previous tasks. The allocation of staff between the subunit manangers had, to some extent, been undertaken with consideration

---

67. Furthermore, the plan outlined the establishment of a computerised information system on technical and economic data which should be fed and used by the subunits and the network of agents. However, the use of the system was delayed for a number of reasons: The information system suffered from lack of compatibility between various computer systems; the organisation members did not understand the way in which the system was supposed to work; and some organisation members lacked the basic training necessary to operate the system. Consequently, simultaneous activities, partially outside the project, were launched in order to deal with these problems.

to the wishes of the single organisation member articulated during a number of manager-employee talks prior to the allocation, but still the single organisation member, whether secretary, technician, salesman or bilingual secretary, tended to react on the new situation by focussing on functional tasks rather than the set of tasks of the team.[68]

In consequence, the project had to deal with a process of implementation that at the stage of initial implementation tended to be fagged out, for primarily two reasons: The organisation members lacked the ability to appreciate that they were part of a team with mutual responsibilities, and above all a responsibility to become aware of the opportunities for developing the product line; and the organisation members experienced a great deal of personal insecurity and responded by focussing on the functional tasks associated with the division of labour prior to the formal reorganisation. These problems related to organisational and human barriers, or resistance to change, as the outcome of a number of reasons for demotivation of the organisation members. *reproducing routines*

The *organisational barriers* related to the level and nature of activities of top management, i.e. directors, regarding the way in which the reorganisation was initiated and followed up during implementation. As mentioned previously, the initiative to the reorganisation came about, partially, in a bottom-up fashion. By accepting the results of the questionnaire and the plan deviced by the human resource director, the top management issued a strong political signal of change. The plan was elaborated by the sales managers, and the ensuing strategy was disseminated through intrafirm conferences. However, the process seemed to stop somewhere between the stages of the formation of attitudes and the initial implementation, for at least four reasons:

(1) At the point of decision, the delegation of tasks was decided upon by the top management, and the subunit managers had to accept the content of this decision without being able to influence the process. *Top down*

(2) During the initial implementation, the subunits were asked to describe the current situation and state of the product line, the threats and opportunities at the various markets, the opportunities for product and market development, and the way in which the subunit would engage in improvements. Each subunit put a lot of work into these descriptions, which resulted in a manual for each subunit, nicknamed cookery books. But more than

---

68. In addition to the segmentation on product lines, a SOS subunit was established with the responsibility of dealing with orders and decisions that had to be made in a hurry. However, the product line subunits tended to regard the SOS activities as tresspassing, and soon the SOS unit had to deal with activities that could not be allocated elsewhere. Furthermore, the SOS unit suffered from a number of interpersonal tensions between its members. These tensions were finally dealt with by replacing one of the members of the unit. The consultants attempted to profile the SOS unit, but at present the unit still seems to be rootless and flottering.

formalizing

five months elapsed from the point of time, where the cookery books were delivered to the top management, to the point of time, where the sales director provided feedback.

(3) Although the single subunit was supposed to be self-contained, the reallocation of the sales personnel did not succeed in achieving this aim. The reason for this was that most of the subunits lacked experience with the product line prior to the reorganisation. As a consequence, the single subunit had to demand and supply each other with man power, and this, primarily informal, interunit exchange was not evenly distributed, thus creating lack of resources at various points of time.

(4) Finally, during the initial implementation, most of the subunit members felt that they needed additional training in the field of planning, allocation of man hours, and detection and solution of organisational problems. However, the top management did not allocate resources for this purpose.

The *human barriers* related to the way in which the single organisation members reacted to the reorganisation of the sales department. Most of them felt insecure and feared that they would not be able to cope with the process and the new challenges that they were exposed to. A wide array of reactions appeared: At the one end, some of the organisation members articulated that they preferred to occupy their old positions; at the other end, some were being over-enthusiastic and expressed delight about the fact that "at last something is happening", no matter what was happening; most of the organisation members were found in the middle, and the single subunit exhibited a large degree of differentiation regarding these reactions. As mentioned earlier, the individual insecurity was primarily handled by focussing on functional tasks and prior training rather than the tasks of the team as a whole. *task environment ?*

The formation of team-like subunits confronted the organisation members with three requirements: (1) You must be able to work in a team, the task environment of which changes more frequently than you are used to; (2) you must contribute to the development of markets and products and thus stimulate the change of your task environment; (3) you have to cross the borders of your usual tasks and participate in the construction of a new organisational role, preferably one that is able to change according to circumstances. Confronted with these requirements, virtually all of the organisation members experienced themselves as insufficient, and they tried to handle their personal insecurity by focussing on the type of tasks that they were good at as a reaction to the fact that the tasks, which they were less good at, began to occupy a predominant space in their self-consciousness.

In firm B, the transition from initiation to implementation was much easier and the type of problems encountered less difficult to overcome. As mentioned previously, firm B had a tradition for strategic planning. The managers of each department were accustomed to device annual plans which were summed up by the board of directors.

However, even though this process was highly routinised and recurrent, the board of directors had some difficulties in coordinating the various plans and projects of each department. The individual departments, and the firm as a whole, focussed on the development of markets, and process and product technology, and at any point of time a large number of projects were initiated and carried out. While this contributed to the competitiveness of the firm, it created, on the other hand, a problem of coordination. The board of directors, the managers, and the employees were unable to hold a total view of the number and nature of projects going on, and especially the amount of resources that these projects demanded, both as demand for man hours within the single department, and as the demand for man hours that the various departments were confronted with by the other departments. In consequence, the resources demanded tended to outgrow the resources available, and in each department there was a tendency to attend to intradepartmental projects at the neglect of extradepartmental demands.

In order to solve this problem, the top management wanted to decentralise the strategic planning and make the lower-level managers and the employees more aware of the amount of resources actually devoted to each project. The organisational solution to this problem was for top management to demand, for each project, a calculation of the man hours required in the department in question and the man hours required from other departments. These calculations would be discussed at interdepartmental meetings which coordinated the amount of resources demanded with the amount of resources available and resulted in a list of priority among projects.

However, this solution was met by two difficulties: First, the process of strategic planning tended to create information overload. Second, while the single departments did seem to function rather well as teams, the members of each department were accustomed to focus on intradepartmental rather than interdepartmental issues, and leave the latter for decision at the board of directors. Thus, they tended to resist the organisational change envisioned by top management.

This resistance was reinforced, because the organisation members lacked the experience and the skills necessary to calculate the demand for resources and the priority of projects in terms of not only intradepartmental, but also extradepartmental available resources. Furthermore, there was a tendency among the organisation members to regard any activity as a project, and there was no clear appreciation of the distinction between project-related and operation-related activities. In consequence, the organisation members tended to believe that there were more resources available for project-related activities than was actually the case.

5.3. Solution: Visibility at three levels

In firm B, the organisation members attended a series of meetings where they were introduced to various methods of organisational and individual planning, and where the vision of the top management was disseminated through discussions and teaching. At the outset, the organisation members were asked to describe the projects that they were currently engaged in and to distinguish between project-related and operation-related activities. The outcome showed a number of more than one houndred projects which eventually shrinked to some fourty-odd projects when operation-related activities were excluded. Next, the organisation members were asked to describe the time duration and the man hour requirements associated with each project in the form of Gantt schemes[69] and make priorities in an ABC-like fashion.[70] As expected, this procedure revealed that the total demand of resources was larger than the amount of available resources, and the process was repeated through interdepartmental discussions. The outcome of the process were the establishment of meetings for coordinating strategic planning, the accept among organisation members of the delegation and decentralisation of strategic planning, and the use of simple planning devices for resource allocation and control.

In firm A, the solution was not equally straight-forward and required a good deal of psychological assistance in order to reveal and remedy the personal insecurity of the organisation members. This was done in a number of meetings between the consultants and the organisation members, individually and in the teams, and the talks with the subordinate organisation members were replicated in the group of subunit managers. The managers suffered, as well, from personal insecurity, but abstained to a greater extent than the subordinate organisation members from discussions with colleagues, since they were more accustomed to work alone and function, literally speaking, as lonely wolves. On the other hand, the level of frustration tended to be higher in the case of the subordinate

---

69. A Gantt scheme, named after Henry L. Gantt (1861-1919), is basically a graphical technique for planning and controlling activities which are depicted on a time scale according to planned and actual duration. The graphical presentation is augmented by a represenation of planned and spent resources, and the activities are divided into two or (preferably) more stages of activity.

70. The ABC-technique is a priority between tasks. Initially, the organisation member describes the purpose of his activities and, preferably, the means by which aims are achieved. Next, the tasks is divided into three groups: (A) The most important tasks which are those tasks that contribute to the development of achievements, and which the organisation member thinks deserve his outmost responsibility; (B) the second most important tasks which are those tasks that, although important, might be undertaken by other organisation members; (C) routine tasks, which are the remaining set of tasks. The ABC-order reflects the order of priority. Any task in the ABC-hierarchy may, at some point in time, become urgent, but the intention is that the ABC-order should be maintained. Finally, resources are devoted to the tasks according to priority.

organisation members who, as subordinates, found it more difficult to voice their complaints and had less opportunities to use the option of exit, i.e. leave the organisation.[71] The psychological assistance aimed at making three types of social behaviour acceptable within the new organisational framework:

(1) As mentioned earlier, because they tended to focus on their own insufficiency, the organisation members concentrated on what they were best at doing and neglected the opportunities for supplementing the work of their colleagues. During the talks, it gradually became acceptable to voice personal insecurity and deficiency of skills, and to ask for and to offer assistance.

(2) Since the reorganisation required that the single organisation member took part in proactive activities and developed their own skills and tasks, it had to become accepted that unit members made suggestions to one another and voiced ideas and visions *vis-á-vis* colleagues and managers. During the talks, this type of behaviour gradually became accepted.

(3) Despite the intrafirm conferences and subsequent subunit meetings, the subordinates felt that the vision of the firm was being blurred by the problems of the daily work life, and they frequently requested management initiatives and direction. An important part of the resolution of individual and organisational tensions was, therefore, to persuade the managers to make strong and clear signals about the direction of organisational activities, and to do this through a dialogue within the group of managers and with the members of their units. The issue of personal leadership was put at the organisational agenda, and the notion of personal leadership was reinterpreted as a responsibility not only of managers, but also of subordinates.

Through the talks, it gradually became accepted that the new organisational configuration required the organisation members to invest themselves, as human beings and colleagues, in a continuos dialogue with one another, where ideas, personal insecurity, sufficient and insufficient competencies were displayed. In sum, the reorganisation in firm A became based on a broad notion of personal leadership exerted at three levels: Within the group of managers; by the individual managers; and within the group of subordinates interacting with their manager. Furthermore, it became accepted that the organisational activities were dependent on the social relationships between the organisation members and required that the organisation members mutually provided psychological security and motivation.

---

71. Notice that during the time of the project, unemployment rates in the Danish economy were much larger than at present.

[handwritten: tendency to fall back on routines to cope with uncertainty]

The way in which these activities were conducted aimed at providing three types of visible organisational behaviour: Personal, strategic, and competence-oriented visibility.[72]

Personal visibility refers to the ability of the organisation members to display their insecurity and insuffiency and to provide assistance for one another through interaction. Organisational innovation redefines the set of organisational roles, and, as the experience from especially firm A showed, most organisation members tend to focus on business as usual in order to avoid the uncertainty incurred by the redefinition of organisational roles. As a consequence, the redefinition of organisational roles becomes blocked, unless something is done to remove the uncertainty at the personal level.

Strategic visibility refers to the responsibility of managers to articulate and communicate the strategic vision of the organisational innovation, and the responsibility of the subordinates to engage in that process. Organisational innovation becomes blocked, unless the routines, which become obsolete, are replaced by a set of new routines that reestablish the means-ends hierarchy. *[handwritten: Obsolete routines must be replaced]*

Competence-oriented visibility refers to the ability of the organisation members to apply their social and professional skills to the new task environment which is created as part of the organisational innovation. In a stable task environment, there exists a balance beween the formal and the informal organisation, and the organisation members are able to combine their social and professional lifes. Organisational innovation disturbs the balance and becomes blocked, unless sufficient resources at the levels of individuals, subunit and organisation are devoted to the development of a new balance between social and professional skills. *[handwritten: new balance in social + professional]*

The organisational innovation in firms A and B exhibit some similarity in the sense that the three types of visibility apply in both cases. Furthermore, in both cases, the organisational innovation was initiated by a performance gap in the sense of a distinct, but not insuperable difference between aspirations and achievements, and thus assumed the character of stress innovation in the sense of March & Simon (1958). However, the way in which the three types of visibility displayed themselves differ between the two cases.

First, as argued previously, the nature of the organisational innovation differ between firm A and B in a way that may be compared to the difference between radical and incremental innovation. Second, while the innovation was initiated by a bottom-up process and decided upon by a top-down process in firm A, it was both initiated and decided upon by a top-down process in firm B. Third, the type of organisational tensions encountered were extremely different, although the top management in both cases tried *[handwritten: top down + bottom up]*

---

72. To some extent, the same objective guided the activities in firm B, although to a smaller degree since the aspect of psychological assistance was much less pronounced than in the case of firm A.

116

[handwritten: ✳ organizational innovation as radical + incremental]

to stimulate a process of change that was deliberately planned and subjected to feedback processes.

5.4. Evidence on the flexibility-stability dilemma

The evaluation of the project in firm A aimed at elucidating the degree to which the expectations of the organisation members had been met by the project, and problems were solved during the process. Initially, the organisation members[73] were presented to the concept of performance gaps and an agenda for subsequent discussion in the subunits. After the discussion, each subunit reported to a plenary assembly, and a dialogue between the subunit teams and the representatives of the board took place. Finally, the conclusions were summed up.

Prior to the discussion, the sales personnel responded to a small questionnaire, and the questionnaire was repeated after the discussion, but before the discussion in the plenary assembly took place. The objective was to detect whether the discussion in the subunits would change the attitudes of the organisation members. The organisation members were asked to evaluate seven questions on a four point scale, and the responses appear from figure 27, where the means of the answers before and after the discussion is depicted. The higher the value, the more the respondents are in agreement with the questions. The difference between the two curves indicate that the discussion in the subunits created a more critical assessment of the change process. Regarding the assessment of expectations, the subunit discussions increased the degree to which the respondents percieved their

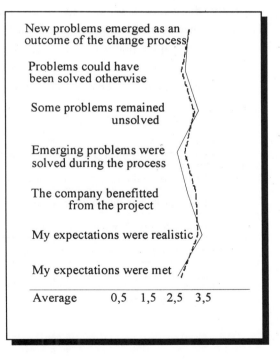

Figure 27. *The means of responses to seven questions in firm A, before (dotted line) and after (full line) the plenary assembly, N=21*

New problems emerged as an outcome of the change process

Problems could have been solved otherwise

Some problems remained unsolved

Emerging problems were solved during the process

The company benefitted from the project

My expectations were realistic

My expectations were met

Average   0,5 1,5 2,5 3,5

---

73. In total, 2/3 of the personnel in the sales department participated in the evaluation together with three representatives of the board of directors.

expectations as realistic and decreased the degree to which they assessed that their expectations had been met. Regarding the solution of problems, the degree to which the respondents assessed that problems could have been solved otherwise, and that some problems were left unsolved, increased, while the degree to which they judged that problems had been solved during the process decreased. Finally, the valuation of the benefits of the project to the firm decreased.

The percentage change of the means are small, cf. figure 28, but seem nevertheless to be valid. In order to support the conclusions, a nonparametric sign test of small samples was employed. The null hypothesis was that the subunit discussion would lower the means, i.e. have a negative impact on the degree of agreement with the questions. The null hypothesis could not be rejected at the 5% level. This result is supported by the increase of the means of the questions whether some problems were left unsolved and whether problems could have been solved otherwise. On the one hand, the increase of the means - contributes to the possibility of rejecting the null hypothesis. On the other hand, the increase of the means reflects a more negative assessment of the change process, and, logically, these means should have been employed in favour of the null hypothesis. However, since the null hypothesis could not be rejected, although the means in question were employed in the opposite direction, the conclusions seem even more valid.

Figure 28. *Percentage changes of the means depicted in figure 27*

Question	%-change
New problems emerged as an outcome of the change process	- 1.61
Problems could have been solved otherwise	+ 5.36
Some problems remained unsolved	+ 4.94
Emerging problems were solved during the process	- 8.62
The company benefitted from the project	- 8.81
My expectations were realistic	+ 5.11
My expectations were met	- 3.99

The cases of firm A and B highlights some of the points made in chapters 2-4. The following three suggestions relate to the human and organisational barriers to change described in section 5.2, while the subsequent three suggestions relate to the opportunities

transitory stage

entailed in the new organisational configuration of firms A and B after the initial problems have been solved as described in section 5.3.

First, as expressed previously in terms of figure 8, the organisational innovation process in firm A tended to stall at the stage of initial implementation which is the transitory stage between initiation and routinisation. Described in terms of figure 17, the rate of change was impeded because of a decrease in job satisfaction associated with the personal insecurity of the organisation members caused by the fact that a congruence between individual goals (as expressed in terms of a balance between social and professional roles) and organisational goals (as expressed in terms of the vision for strategic planning) had not been established. *liason relations*

Second, both cases reflect the problem of uncertainty in the Galbraithian sense of an information gap. In firm B, top management faced the task of reconciling departmental strategic plans without being able to judge among the number of projects and thus make priorities. In consequence, through additions to departmental organisational repertoires, discretion moved closer to the points of origin of information, and liason relations were established in the form of interdepartmental meetings. In firm A, the sales personnel felt unable to appreciate the product technology and development opportunities of the firm, and this perception of a performance gap had gradually increased organisational tensions to a degree, where initiation was undertaken by a director who was not even directly linked to the activities of the sales department. An attempt to resolve the organisational tensions and increase the information processing capacity of the sales department was undertaken by establishing self-contained task environments in subunits. This attempt was supported by the creation of a vertical information system, a buffering mechanism (the SOS unit), and the establishment of a new management position supposed to coordinate the subunit managers and report to the sales director.

Third, however, the case of firm A shows the difficulties entailed in the transition from initiation to implementation. The organisational innovation resulted in the creation of more complex attention centres, which in terms of figure 17 ought to facilitate the implementation of the organisational innovation. Furthermore, the team-like nature of these attention centres decreased formalisation and centralisation, and, furthermore, contributed to reducing the degree of stratification and the number of subgoals, since various specialties were brought together in self-contained task environments. In theory, these achievements ought to support the implementation of the organisational innovation. However, this effect was difficult to realise in practice, as described in section 5.2. This may be ascribed to a diversity of attitudes, to insufficient feedback, and to a lack of power of innovation champions. Regarding the diversity of attitudes, the effectiveness of the self-contained task environment tended to be traded off by the focus of organisation members on functional tasks. Regarding the insufficient feedback, the prolonged period of feedback from top management on the cookery books decreased the motivation of the organisation members to participate in the implementation process. Regarding the lack of power of innovation champions, the subunit managers were supposed to push the implementation process through a dialogue with their subordinate, but they had no influence on the delegation of tasks at the decision substage. Furthermore, the transition from initiation to implementation was complicated by the fact that the degree of self-containment of tasks was lowered by the interunit demand on resources described in section 5.2.

Later, the transition from initiation to implementation was further complicated by a development not described in sections 5.2-5.3. During the last two months prior to the evaluation, firm A experienced a rapid increase in incoming orders. This created time pressures across the unit staffs, and the organisation members began to voice complaints about the decrease of time which they could devote to integrating the intraunit activities. As described in chapter 3, cf. figure 11, time pressures decrease the span of attention.

Fourth, in terms of figures 11 and 14, the creation of self-contained tasks increases the tolerance for interdependence. Moving discretion closer to the points of origin of information increases the ability of decision makers to predict outcomes and thus judge between good and poor outcomes. On the other hand, coordination within the units relies increasingly on feedback communication. Coordination by feedback is less efficient than coordination by plan unless the feedback process is efficient. However, feedback may become efficient as the organisation members learn to interpret and distribute feedback information. In the case of firm A, the initiatives described in section 5.3 are conducive to the process of feedback. From an action learning perspective[74], Gay (1983) describes a number of instruments which may facilitate organisational change, such as: The improvement of management processes by developing the skills of subordinates; the recognition among managers that their role as a manager is not solely dependent on their

---

74. Briefly put, action learning may be defined as the interactive process of a number of organisation members who try to solve current problems through joint reflection and reinterpretation (Garratt, 1983). Action learning aims at the resolution of organisational conflicts through reinterpretation of past experience in a problem-oriented fashion (Revans, 1983). In its ideal form, an action learning programme attends to a crucial organisational problem, involves organisation members who are willing to develop their own skills, allocates authority to these members to take action on the problem, and involves procedures for reflective learning (Garratt, 1983, pp.31-32). Reflective learning may be described as learning about learning in the sense of Argyris & Schon (1978), i.e. the organisation members evaluate the previous contexts for learning. The notion that organisation members interactively become able to solve immediate problems and learn from the problem solving process has been described as the basic idea of the *learning company* by Pedler, Burgoyne & Boydell (1991).

own activities, but is influenced by the adjustment or expansion of the roles of other managers; and the recognition of managers that they should initiate changes in their own prescribed role. All of these aspects were present in the development of an attitude favourable to personal leadership described in section 5.3. In the long run, the promotion of personal leadership may facilitate double-loop learning in the sense of Argyris & Schon (1978), i.e. learning may not only involve the process of assessing the task environment, compare this assessment to performance programs, and initiate appropriate action on problems, but also the process of assessing the usefulness of the performance programs as such. As emphasised by Garratt (1983), this involves the ability of the organisation to reinterpret deviations from operational plans and budgets.

Fifth, as the subunits in firm A learn to function in a team-like fashion, and as the departmental organisation members in firm B become accustomed to use their planning devices for interdepartmental coordination, the ability of the organisations to resolve conflicts may increase, since the distance between the perceptions of the interacting organisation members becomes smaller. However, in the case of firm A, there is a potential risk for the development of an increasing diversity of rationalities, in two dimensions. The first dimension relates to the intradepartmental activities of the sales department. In terms of figure 11, as each subunit develops its own task environment, subgoals may become differentiated, and the persistence of subgoals may increase. In effect, the subunit focusses of attention may become more dispersed. The second dimension relates to the interdepartmental relationship between the sales and production departments. As the sales department subunits develop their competencies and become more engaged in the development of products and markets, the rate of innovation proposals may increase. However, the production department is still organised through functional lines, and during the evaluation the managing director expressed some concern about the ability of the production department to keep up with the sales department. He anticipated a future need for human resource development in the production department, but was unable to assess whether the firm had the resources necessary for the task.

The production department problem of keeping up with the sales department may result in organisational tensions for the following reasons: The establishment of self-contained tasks in firm A implies that the interdependence between organisation members tends to become reciprocal, while the interdependence between units tends to become pooled in the sense of figure 23. According to Thompson (1967), pooled interdependence is associated with low frequency of decisions, less importance of contingencies, and a small amount of communication. To the extent that innovation proposals occur in the sales department, the interdependence between the individual sales subunit and the production department tends to become reciprocal at the stage of the initiation of product innovations.

developing competences & promoting engagement of innovation proposals

121

However, to the extent that the production department is confronted with different requirements from the sales subunits, the burden of managerial tasks at the level of the linking sales department manager and the sales director may increase, thus creating an information overload.

Sixth, an important aspect of the organisational innovation in firms A and B is the stimuli to search provided by the changes. In firm B, the addition of new routines to the repertoire of planning stimulates reproductive search in the sense that the organisation member is required to scan the task environment for additional information relevant to planning. In firm A, the work on the cookery books relied on reproductive search, and the objective of the new subunits is to stimulate opportunistic surveillance. However, again one may anticipate future problems in firm A. To the extent that the cookery books contribute to performance, firm A will find itself at position B in figure 12, and aspiration levels will tend to increase. To the extent that the aspiration levels of the sales department exceeds the aspiration levels of the production department, organisational tensions at the horisontal level may occur. Furthermore, to the extent that the board of directors temporise their response to proposals from the sales department, as in the case of the cookery books, organisational tensions at the vertical level may occur.

new routines in planning
stimulates reproductive search
opportunistic surveillance

Chapter 6
Learning Cycles and Some Concepts

[handwritten: threats + obstacles]
[handwritten: recombinant knowledge ✳]
[handwritten: + new knowledge]

6.1. The incompleteness of learning cycles

The processes of decision making based on search and integration of new knowledge in the organisational repertoire, which have been described in chapters 2-4, focussed on a number of threats and obstacles to organisational effectiveness and the way in which these problems may be solved. Chapter 5 described two cases of organisational innovation which presented various ingredients to the solving of problems related to the implementation of organisational innovation, based on action learning. As chapter 5 showed, the processes of decision making tended to be ambigous, and the implementation of the organisational innovation created a number of potential problems which may impede organisational performance, result in future performance gaps, and stimulate learning.

The type of learning involved in the two cases may be described as a combination of (1) the modification and recombination of the knowledge present in the organisation and (2) the creation of new knowledge. Especially in the case of firm A, this combination aimed at developing the knowledge of organisation members and units, the long term objective of which was to enhance the ability of the organisation as a whole to learn about threats and opportunities, interpret the new knowledge and use this interpretation to develop products and markets. Thus, following Hedberg (1981), the nature of learning involved in the organisational innovations may be described as adaptive-manipulative, i.e. in terms of processes "whereby learners iteratively map their environments and use their maps to alter their environments" (ibid., p.4). *[handwritten: Mapping environment]*

Hedberg (1981) suggests that it is meaningful to use the concept of *organisational* learning, although only organisation members, not organisations, are able to learn. The learning which takes place within the social context of the organisation is, of course, done by the organisation members, i.e. the human beings who inhabit the organisation. However, the outcome of their learning is more than a cumulative result.[75] Rules and procedures are the manifestations of previous processes of learning, and they impose behavioural regularities on the organisation members. Organisations have ways of storing

[handwritten: ← rules + procedures represent past learning ✳]

---

75. This proposition is parallel to the Nelson & Winter (1982) argument that the organisational repertoire is something more than the cumulative result of the individual skills of the organisation members, and that individual skills only are meaningful in a social context, cf. section 1.3. Hedberg (1981, p.6) argues that "organizations influence their members' learning, and they retain the sediments of past learning after the original learners have left".

storing information artifacts & behaviors

information, not only in data banks and accounting figures, but also in the procedures which define how data are collected, processed and stored. Organisations develop basic assumptions about organisational activities and the way in which these activities relate to the extraorganisational environment.[76] As described in chapters 3-4, the organisational configuration defines the mechanisms whereby organisation members attend to problems and problem solving.

The discussions in chapters 2-5 showed that it is useful to describe problem solving in an organisational context as a complete cycle in the sense that the perception of performance gaps trigger a process of search which leads to a problem solution that sets the stage for future performance gaps. However, chapters 2-5, especially the latter, did also show that there exists a number of incidents at which the cycle becomes broken, e.g. due to code scheme barriers, resistance to change, intraorganisational political struggles, information gaps and the like. In consequence, the discussion in chapters 2-5 implies that two questions are of importance to the understanding of organisational learning: What are the mechanisms by which organisational learning may take place as a complete cycle? For which reasons are the complete cycle broken, and which types of learning are associated with these breaks?

Impact of broken cycles

Hedberg (1981) has elaborated on these issues in a two-step argument: First, the complete cycle of learning may be described in terms of a stimuli-response model which implies that organisations are able to accommodate to environmental changes and feed back stimuli to the environment in consequence of the changes of the organisational behaviour. Second, the complete learning cycle is broken because there is no direct relationship between environmental stimuli and organisational response, and between organisational stimuli and environmental response. _Stimulus response model_

The stimuli-response model described by Hedberg (1981) implies the following type of organisational learning when organisations interact with their environment: Changes in the extraorganisational environment provides stimuli for organisational action. The organisational repertoire, which determines organisational action, may be described in two dimensions, i.e. in terms of a set of rules about how to interpret the environmental stimuli, and a set of rules about how to response to organisational stimuli. The interpretation of environmental stimuli takes place according to the beliefs, or theory of action (Argyris & Schön, 1978), which determines how the organisation, as a whole, percieves stimuli. The

---

organization culture values artefacts

76. Schein (1985) argues that organisations have *cultures* which are embodied in the *basic assumptions* on the relationship between the organisation, the organisation members, and the extraorganisational environment; *values* about action which is approved socially; and *artifacts* in terms of e.g. technology and the visible behaviour of organisation members.

theory of action Δs (handwritten annotation)

theory of action changes as feedback on the assessment of action outcomes diverts from the initial beliefs. While there is a tendency towards gradual change in the rules which define how responses are assembled, there is a tendency towards more discrete changes in the rules which define how stimuli are interpreted. This difference comes about because it is easier to modify decision rules than to change the theories of action which determine the decision rules. *decision rules* (handwritten annotation)

In consequence, "organisational changes are constrained to shifts inside behavioral modes" (Hedberg, 1981, p.9), and Hedberg (1981) suggests that organisational changes within a complete cycle of learning may be described by three modal situations: (1) *Adjustment learning*, which refers to the change of parameters or rules. Adjustment learning is adequate in situations where the relationship between the organisation and its environment is subject to minor changes. The change of parameters or rules are achieved relatively easy, may even be routinised, and results in fast responses. (2) *Turnover learning*, which refers to the change of the organisational repertoire by adding new routines and removing some of the old ones, i.e. the set of rules of interpretation and response assembly changes. Turnover learning occurs when the relationship between the organisation and its environment is subject to significant changes. The change of the organisational repertoire is difficult and its duration exceeds the short run. (3) *Turnaround learning*, which refers to changes of the theory of action. Turnaround learning is necessary when the relationship between the organisation and its environment changes substantially, but it is, of course, extremely difficult. When it occurs, the nature of organisational action changes rapidly. *What about learning by doing, etc* (handwritten annotation)

The direct stimuli-response links entailed in the relationship between the organisation and its environment may be broken for a number of reasons, which may be described by reference to the model of rational adaptation presented in chapter 3. March & Olsen (1976) have argued that even if we retain the assumption that organisational behaviour is adapted in terms of experience, the process of adaptation may not be rational, since adaptation requires the interpretation of experience which often takes place in conditions of ambiguous or conflicting goals. The phenomenon of conflicting goals, which entered the discussions of chapters 2 and 4-5, is associated with the beliefs and intentions of the organisation members. These beliefs and intentions affect the way in which experience is interpreted, since interpretation requires an image of the problem world, i.e. a model of thought (or theory of action). This model of thought often implies the impression[77] that

---

77. Both among researchers and organisation members. Please notice that March & Olsen (1976) are occupied with the requirements of an adequate theory of organisational behaviour. However, the familiar theoretical conceptions of organisational behaviour are often reflected in the way in

organisational events occur as a result of intentions, which, however, may not be true: Organisational behaviour tends to be guided by rules which have little impact on individual intentions; individual action may be less connected to intentions than to organisational roles; and the decision making process is often overwhelmed by exogenous factors (March & Olsen, 1976, pp.19-20).

In consequence, March & Olsen (1976) question the validity of the complete cycle of choice (cf. figure 29) which they argue often enters the conception of organisational behaviour. The complete cycle of choice is closed and based on four connections:

(1) The cognitions and preferences held by individuals affect their behavior.
(2) The behavior (including participation) of individuals affects organizational choices.
(3) Organizational choices affect environmental acts (responses).
(4) Environmental acts affect individual cognitions and preferences.

(March & Olsen, 1976, p.13)

Figure 29. *The complete cycle of choice*

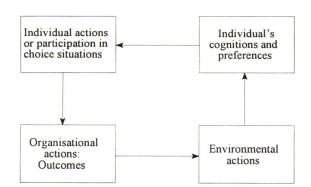

Source: March & Olsen (1976, p.13), figure 1.0

March & Olsen (1976) propose the following argument:

First, the relationship between individual cognitions and preferences, and individual action, is ambiguous. The individual inclination to action depends on resources and time,

---

which organisation members understand organisational behaviour, especially because a large part of organisation theory is normative and reflects the praxis of organisations.

and the individual capacity for action is often smaller than the capacity for preferences and cognitions. The opportunity to act presents itself in a *set* of choice situations, and not only in a single situation. Furthermore, individual action depends not only on preferences or cognitions, but also on organisational obligations, and most of the individual action in an organisational context is guided by rules. Finally, individual cognitions and preferences are not exogenous to action, and "beliefs and preferences appear to be the result of behavior as much as they are the determinants of it" (March & Olsen, 1976, p.15).

Second, the relationship between individual and organisational action is not unidirectional. Organisational action cannot simply be derived from individual action, since we are dealing with a social system where the decision making process serves other ends than organisational action, such as the maintenance of the organisation as a social system through processes of creating legitimacy and preserving interpersonal relations.

Third, environmental acts are not only responses to organisational action. The organisation is but one constituent part of the environment, and environmental action comprises the interrelations between a set of actors which interact on the basis of given structures and occurring extraorganisational events.

Fourth, individual preferences and cognitions may only to a moderate extent be affected by environmental acts. Environmental acts are often not percieved by the organisation members, and when they are, they do not present themselves to the individual as objective data, but as events which require interpretation. Interpretation is based on the existing set of perceptions and cognitions, is subjected to myth and ideology, and often relates to events which have been observed by others than those who interpret the significance and content of these events.

6.2. The neo-behavioural approach

The qualifications to the closed cycle of choice reflects the point of view that an organisation

> is a set of procedures for augmentation and interpretation as well as for solving problems and making decisions. A choice situation is a meeting place for issues and feelings looking for decision situations in which they may be aired, solutions looking for issues to which they may be an answer, and participants looking for problems or pleasure.
>
> (Cohen, March & Olsen, 1976, p.25)

Thus, decision making is not only triggered by the occurrence of problems, but also by a propensity of organisation members to generate choice situations. Furthermore, for the reasons argued above, the occurrence of problems is difficult to percieve by the organisation members, since individual preferences and the relevance of occurring problems are ambiguous, and since preferences "are discovered through action as much as being the basis of action" (Cohen, March & Olsen, 1976, p.25). Instead of a process of rational adaptation, Cohen, March & Olsen (1976) suggest the metaphor of a *garbage can*

> into which various problems and solutions are dumpted by participants. The mix of garbage in a single can depends partly on the labels attached to the alternative cans; but it also depends on what garbage is being produced at the moment, on the mix of cans available, and on the speed with which garbage is collected and removed from the scene.
>
> (Cohen, March & Olsen, 1976, p.26)

The provocative metaphor of a garbage can implies that the decision process is viewed as a number of "streams", such as (1) *problems* which are attended to - or, more precisely, are present in the organisation, since problems and decisions are not, necessarily, interrelated; (2) *solutions* which are produced, but not in a straight-forward manner since solutions do not, always, presuppose a question, but are often found in advance and serve to raise questions that make the solution relevant; (3) *participants* who engage themselves in a decision making process, but in a discriminatory manner since their attention is divided among several demands; and (4) *choice opportunities* which emanates as occasions on which the organisation is required to produce a choice in relation to a problem. The interaction between these "streams" depends on the time patterns of, and the individual and organisational attention devoted to, problems, solutions and choices, and is characterised by the frequent decoupling of problems and choices: Solutions to problems may be discharged as new solutions appear, although this continued process does not resolve the problems in questions; solutions which define some problems may be implemented without attention to existing problems; and the organisation may be aware of some problems but leave them be in order to attend to current activities (Cohen, March & Olsen, 1976, pp.33-35). In consequence, decision making is often not directly linked to problems, but problems "are worked upon in the context of some choice, but choices are made only when the shifting combinations of problems, solutions, and decision makers happen to make action possible" (ibid., p.36).[78]

---

78. Cohen, March & Olsen (1976) describe a number of decision situations by simulating organisational decision making of two types: Participation and access, where participation refers to the opportunity of decision makers to participate in choices, and access refers to the opportunity of problems or solutions to align with choice opportunities. Participation may be *unsegmented* in the sense that each decision maker is allowed to enter any decision situation; *hierarchial* in the sense that the entry to decision situations is gradually limited as we move down the hierarchy; and *specialised* in the sense that decision makers attend to only one choice, and each

The garbage can model provides an alternative to the model of rational choice in conditions when the assumptions of that model is not met (which is frequently the case, as argued above). Cohen, March & Olsen (1976) recognise that the decision making processes of the garbage can type

> does not do a particularly good job of resolving problems. But it does enable choices to be made and problems sometimes to be resolved even when the organisation is plagued with goal ambiguity and conflict, with poorly understood problems that wander in and out of the system, with a variable environment, and with decision makers who may have other things on their minds. This is no mean achievement.
>
> (Cohen, March & Olsen, 1976, p.37)

The decoupling of problems and choices result in interferences at each of the four arrows in figure 29, which from the point of view of rational decision making may be regarded as pathological. In the absence of interferences, individual beliefs and attitudes are translated into individual actions which are transformed into organisational action through an organisational decision process. The economic environment of the organisation, e.g. competitors and consumers, react on the basis of a repertoire of responses that are observed and interpreted by the organisation members according to their theories of the world, and form the basis for an adaptation of beliefs and attitudes which are translated into individual action *etcetera*.[79] However, interferences occur since organisations, as argued previously, operate under conditions of ambiguity in relation to the links between (1) preferences and actions, (2) individual behaviour and organisational choice, (3) organisational action and environmental response, and (4) environmental acts and individual preferences.

March & Olsen (1976a) describe these interferences in terms of *experiential learning* which refers to the type of learning that occurs under conditions of ambiguous or conflicting goals where the experience of past actions requires interpretation. Experiental learning may be associated with one or more of the following "pathologies", cf. figure 30:

Role-constrained learning appears as a break of the complete cycle at the link between individual beliefs and action. Interpretation is not translated into individual action, since individual behaviour is confined within the set of organisational routines which manifest

---

choice is associated with only one decision maker. Similarly, access may be unsegmented, hierarchial and specialised. They argue that unsegmented cases are rare, since problems "come to seek connections to choice-opportunities that solve them, solutions come to seek problems they handle successfully" (ibid., p.31).

79. In this sense, the complete feedback cycle may cause both single-loop and double-loop learning, as described in section 5.4.

Figure 30. *The incomplete learning cycle*

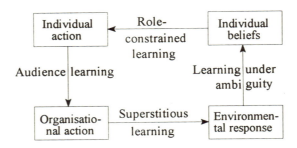

Source: March & Olsen (1976a, pp.56-59)

themselves in organisational roles and standard operating procedures. Although individual beliefs may become manifest in individual action, individual and organisational action may be decoupled. This is the case of *audience learning* where the outcome of individual learning has no effect on organisational action because the organisation member is unable to influence the way in which the organisation attends to and solve problems, e.g. since organisational action as a reflection of a social system may serve other purposes, as argued in section 6.1. If individual beliefs become manifest in individual action which translates into organisational action, the result may reflect a perverse relationship between organisational action and environmental response, because the organisation members fail to interpret the environmental responses, or interpret them erroneously. This is the case of *superstitious learning* where organisational behaviour has but a small effect on environmental responses, although the organisational action was based on an interpreta-tion of the extraorganisational consequences. Finally, even if interferences are absent at the links between individual beliefs, individual action and organisational behaviour, the organisation members may be unable to interpret subsequent events. This is the case of *learning under ambiguity* where

there exists no single, objective explanation of an outcome or of its causes. Different individuals may interpret situations and processes differently because of excessive problem complexity, selective perception, different cognitive styles, and mental maps.

(Hedberg, 1981, p.11)

130

In conclusion, although no references to the concepts of bounded rationality and satisfying behaviour have been made in this section, the idea that organisation members act within computational and cognitive limits in order to produce acceptable results is present in the neo-behavioural theory. In a number of respects, the neo-behavioural theory presents a more broad version of these concepts. The following six arguments may validate this point of view by comparing the neo-behavioural setting with the behavioural one.

Rational decision making

First, decision making is still rational in the sense that the organisation members interpret events, form judgements, and try to act accordingly. However, this process takes place within a framework of ambiguity where the focus of attention to a greater extent than in the behavioural theory is moulded by social interaction. The organisational choice situation becomes an arena where organisational "streams" form and break waters with the result that misrepresentations or non-verified theories of the intra- and extraorganisational environment may persist as long as the organisation survive economically and the organisation members are able to make sense of what they are doing.

Second, although it is acknowledged that individual behaviour in an organisational setting is rule-guided, the rules of the game are less sequential and may to a larger extent produce results which were intended by nobody.

Rules of the game

Third, even though organisational learning in many instances is determined by problemistic search, the problems to which the organisation members attend are less factual since they often appears as cognitive representations of what the organisation members think they see rather than what they more or less objectively may assess. Furthermore, the phenomenon of opportunistic surveillance occur more frequently in a neo-behavioural setting, reflected in the appearance of solutions to which no current problems are attached.

learning determined by problemistic search

Fourth, the learning potential of the organisation may be greater than percieved by the behavioural theory due to the incidence of audience learning which enriches the skills and knowledge of the organisation member, but does not translate into organisational action. On the other hand, one may argue that the incidence of audience learning implies that the effective learning of the organisation, i.e. learning which results in organisational action, may appear at a lower rate in a neo-behavioural than in a behavioural setting.

Fifth, while the behavioural theory to an important extent is preoccupied with the dysfunctions of bureaucracy and presents a normative prescription for the resolution of these dysfunctions, the neo-behavioural theory goes one step further and argues that these dysfunctions are inevitable; however, rational decision making is still an option, and garbage can decision processes produce organisational choices which are able to resolve conflicts and secure the long term survival of the organisation.

behavioural theory & dysfunction of bureaucracy

Sixth, the interaction between organisation members reinforces the context of individual behaviour on the basis of cognitive consistency. Organisation members tend to believe that their formation of attitudes and their mental mapping of means-ends relationships are interdependent, and to the extent that this belief is reflected in individual action, behavioural regularities result. This may be explained in terms of the extent to which the organisation member is integrated into the organisation or alienated from it. March & Olsen (1976a, pp.63-65) argue that to the extent that the organisation member is integrated in the organisation, he will identify with the organisational events which he observes; and conversely if he is alienated from the organisation. To the extent that the organisation member trust the perceptions of the organisation members with whom he interact, he will share their preferences and views on organisational events; and conversely in the case of distrust. These two basic mechanisms form the core of experiental learning, or one may even use the phrase *contextual learning* since this formation of beliefs, interpretations and subsequent translation into individual action is based on social interaction.

6.3. An evolutionary approach

The behavioural reasoning described in chapter 3 has served as an inspiration to evolutionary theorising, especially in the work by Nelson & Winter (1982) who describe the ideas of bounded rationality and satisficing behaviour as one of the allies or antecedents of evolutionary economics:

> Our basic critique of orthodoxy is connected with the bounded rationality problem. We base our modeling on the proposition that in the short and medium run the behavior of firms can be explained in terms of relatively simple decision rules and procedures. Much more than the behavioralists, however, our concern has been with economic change. Therefore, we have put much more stress than they on processes that link changes in firm decision rules and procedures (including productive techniques) to a changing economic environment.
>
> (Nelson & Winter, 1982, p.36)

The Nelson & Winter (1982) approach presupposes rational decision making, but differs from the behavioural approach in the sense that mechanisms for variation and selection to a greater extent are allowed to interact with the mechanisms that preserve and transfer organisational knowledge. Furthermore, while the behavioural approach focusses on intraorganisational matters, the Nelson & Winter (1982) approach focusses on interorganisational dynamics in an open system framework. As described by Knudsen (1991, pp. 315-16), the evolutionary models presented by Nelson & Winter (1982) are based on three

[handwritten: Expression?]

[handwritten: Reproduction,]

mechanisms: *Transmission*, described by the existence of routines which transfer and reproduce the relevant knowledge of the organisation; *variation*, described by the occurrence of innovations, as the analogy to mutations in biological models; and *selection*, described as the discrimination of the economic environment between routines. Nelson (1987, pp.12-13) describes variation as "a mechanism that introduces novelties to the system" and selection as "some understandable mechanism that 'selects on' entities". In relation to the use of the notion of selection for "deriving hypotheses in evolutionary economics", Witt (1992) argues that a number of prerequisites must be fulfilled:

[handwritten: → argument against evolutionary hypothesis]

> First, competition, i.e. selective pressure, must be fierce enough to cause differential survival and/or growth rates of the economic decision units. (...) Second, there must be sufficient variety of economic behaviour subjected to selection. (...) Third, there is a possibility that, under certain conditions, heterogenous (e.g. poor or superior) performers do better in the presence of one another. Then they coexist in the selection or population equilibrium (polymorphism). (...) Fourthly, the variety subjected to selection pressure must not induce competing selection tendencies. (...) Fifthly, and perhaps most importantly, there must be sufficient time for the selection process to operate on variety. (...) If the translation of novelty into innovative economic activities happen at a very high pace, there may be hardly any inertia which selection could systematically amplify differentially.

[handwritten: variety competition in selection]

[handwritten: TIME]

[handwritten: novelties to the system]

(Witt, 1992, p.15)

[handwritten: Now, look in R - 12]

The main focus of this argument by Witt is on the interplay between the mechanisms of transmission, variation and selection in a longitudinal perspective. In order to make the system work, the mechanism of variation must be confined within upper and lower bounds consistent with the lower bound of the selection mechanism and allowing inertia associated with the persistence of at least a major part of the set of existing and available routines. If these requirements are fulfilled, the system may work in a way which creates regularities of behaviour at the macro level, and thus comprises features that might be described as properties of an equilibrium position. However, these regularities might easily be determined by organisational behaviour, the observation of which would lead the mind of the researcher far astray from the path of equilibrium thinking. This is, for instance, the case of the model presented by Chiaromonte & Dosi (1993), where

[handwritten: aggregate variables]

> the possible regularities in the dynamics of aggregate variables (such as income and productivity) are *emergent properties* of a system that self-organizes far from equilibrium.

(Chiaromonte & Dosi, 1993, p.108)

The non-equilibrium micro level properties in the Chiaromonte & Dosi (1993) model, which may constitute macro level regularities, are reflected in phenomena such as the endogenous nature of technical advance, the coevolution of routines and the economic environment, and the differences between firms in relation to their ability and committ-ment to innovation.[80] The ability of the firm to engage in innovative activities depends on the past experiences in this direction since learning are cumulative, both due to the cumulativeness of intraorganisational learning and to the occurrence of learning externalities[81] (ibid., pp.109-10). *learning outcomes*

The outcome of learning is embodied in the individual and organisational routines. In the Nelson & Winter (1982) approach, routines are the economic analogy to the biological notion of genes and refer to predictable regularities of behaviour. They appear as decision rules which determine the type of action appropriate in given circumstances. Routines are embodied in the physical outline of productive activities, individual skills and decision rules. At the level of the organisation member, "individual skills are the analogue of organizational routines" (ibid., p.73) and form the repertoire of the organisation member (ibid., p.98). By skill, Nelson & Winter (1982, p.73) mean "a capability for a smooth sequence of coordinated behavior that is ordinarily effective relative to its objective, given the context in which it normally occur". The organisational repertoire consists of the collection of individual repertoires, but organisational action becomes only effective to the extent that the individual repertoires are coordinated in productive activities. Similarly, the knowledge possessed by an organisation is the collection of individual knowledge, which, however, "is meaningful and effective only in some context, and for knowledge exercised in an organizational role that context is an organizational context" (ibid., p.105). The organisational knowledge is produced through "shared experiences in the past, experiences that have established the extremely detailed and specific communication system that underlines routine performance" (ibid.). The performance of routines is sensitive to the frequency by which the routines are invoked in action, and consequently Nelson & Winter (1982) argue that organisational coordination and knowledge is preserved through "remembering by doing" (ibid., p.107).

remembering by doing

coordinated repertoires *skills = routines* ** definition of a skill*

80. Mostly, the evolutionary work referred to in this study takes on the character of population thinking in the sense that the object of the study is the outcome of the interplay between competing organisations which employ different sets of decision rules. Thus, the evolutionary work in question may be defined as interorganisational theorising, as mentioned above, and analysis of intraorganisational properties which are subjected to selection. The present section aims at describing these intraorganisational properties, i.e. to derive from the interorganisational work in question a set of intraorganisational learning characteristics which can be employed in the subsequent sections 6.4-6.5. These characteristics serve as an important set of microeconomic assumptions and properties essential to the working of the interorganisational evolutionary models.

81. Chapter 7 returns to this issue.

performance gaps change based on the accountability system & external environment, but historical transmission of routines

tacit knowledge
in education

Inspired by Polanyi (1962), Nelson & Winter (1982, pp.76-82) argue that knowledge on how to perform organisational tasks is, to an important extent, tacit because there are limits to the articulation of knowledge among organisation members. These limits relate to the amount of information necessary in the communication of knowledge. First, the organisation member does not possess the computational capability necessary for attending to all the details of the performance of a certain operation or skill. Second, even if the organisation member was in possession of the required computational capability, the amount of time necessary for the communication of all details would be excessive. Third, the individual organisation member is unaware of many of the details entailed in his performance of certain activities and thus unable to communicate these details. Fourth, even if these constraints were absent, "the linear character of language-based communication" (ibid., p.81) would create an incoherent message. In consequence,

reason tacit

much operational knowledge remains tacit because it cannot be articulated fast enough, because it is impossible to articulate all that is necessary to a successful performance, and because language cannot simultaneously serve to describe relationships and characterise the things related.

formal + tacit in petitions   (Nelson & Winter, 1982, pp.81-82)

In the Nelson & Winter (1982) setting, the ability of the firm to deal with performance gaps may be described by three defining characteristics: *First,* the repertoire of routines serve as an internal selection environment that discriminates between solutions feasible to the firm. The internal selection environment is dynamic in the sense that it changes as the outcome of learning. *Second,* the context in which organisational action takes place serves as an external selection environment that determines whether the solutions feasible to the organisation are adequate in order to satisfy the goals invoked by the organisational action. Like the internal selection environment, the external selection environment is dynamic in the sense that it changes as the outcome of learning processes in competing organisations that impose performance gaps on each other. *Third,* there is an interactive relationship between the internal and external selection environment in the sense that the external selection environment defines the limits of survival of the internal selection environment, while the internal selection environment affects the properties of the external selection environment by introducing new productive solutions.

not fixed
not fixed
how clusters cope w/ petition

Essential to the ability of the organisation to cope with performance gaps is the nature of the mechanism of transmission which transfers and reproduces the relevant knowledge. The relevant knowledge refers to the knowledge which the organisation applies in order to close performance gaps, and its application is guided by routines which, as indicated above, is the manifestation of the organisational history. Nelson & Winter (1982)

dealing w/ performance gaps
① repertoire of routines - internal selection
② context, external selection environment
③ internal/external interactive relationship

[handwritten top margin: 3 types of routines look like ? What do these in education.]

distinguish between three types of routines, to some extent characterised by different time horizons: *Operating procedures* which govern short run behaviour and reflect the existing factors of production; *investment rules* that govern the period-to-period change of "those factors of production that are fixed in the short run" (ibid., pp.16-17); and *search procedures* which are routines for changing, over time, the routines employed in the short and medium run.[82] The outcome of organisational action determines the degree of persistence of these routines. On the one hand, if the outcome is preferable, the propensity of the firm to employ precisely those routines which proved valuable increases. This argument was previously presented in chapter 3 by reference to Cyert & March (1963). On the other hand, the survival and thus persistence of these routines mould, gradually, the external selection environment as the organisation expands its activities, and thus contribute to an extraorganisational environment favourable to the surviving routines.[83]

The workings of the different types of routines described by Nelson & Winter (1982) may be illustrated by the Nelson & Winter (1974) model on long run productivity growth[84]: Firms are supposed to adhere to their standard operating procedures associated with the stock of capital in the sense that the decision to invest is not only determined by an investment rule, but also by a rule that governs the choice of what sort of technique should be bought; this choice-rule, which is a routine for changing routines, determines that the investment decision takes, as its point of departure, the existing stock of capital; search for new techniques is local in the behavioural sense, i.e. the firm survey and monitor possible techniques similar to those already in use; in consequence, the standard

[handwritten: molding the external selection environment]

82. Nelson & Winter (1982) refer to the description by Cyert & March (1963) of a hierarchy of procedures, where higher-order procedures are employed to change lower-order procedures.

83. In consequence, if this relationship between the internal and external selection environment was the only type of dynamics at work, the evolutionary system would settle down, in the long run, on a stationary state, e.g. a stable growth pattern where the surviving firms would continue to exist indefinitely. This argument is inspired by Cross (1983) who is preoccupied with the convergence of systems of adaptive economic behaviour towards equilibrium. However, as described in chapters 7-8, the system may comprise coexisting features which cause the economic performance of the system and the single organisation to decline and subsequently improve, according to the nature of the adaptive process in question. Furthermore, even if some organisations do persist and preserve a dominant market share for a very long period of time, as shown by scholars in economic history, this is not due to a convergence towards a stable pattern, but to the ability of these organisations to exploit their capabilities, e.g. in order to innovate, move quickly down the learning curve, secure the procurement and development of vital inputs, and establish distribution channels (Chandler, 1992).

84. A revised version of the model is presented in Nelson & Winter (1982, chapters 8-9), and Nelson (1987) employs the model in his presentation of evolutionary modelling.

standard operating procedure
defines search agenda

operating procedures define the agenda for search and thus provide the organisation with a stable pattern of evolution of routines.

Even in the absence of investments, i.e. when the decision rule on the choice of technique is not invoked, a gradual change of operating procedures is likely to occur. This is the case of *learning by doing* in the sense of Arrow (1962) who argues that learning is "the product of experience" and "only take place through the attempt to solve a problem and therefore only take place during activity" (ibid., p.155). He associates learning with a process of repetition subject to increasing returns and assumes that learning, described in terms of the accumulation of experience, may be measured by cumulative gross investment, reflecting the proposition that knowledge (technical change) is capital-embodied. Arrow (1962) refers to, among others, Lundberg (1961) who describes the *Horndahl effect*, i.e. the experience of a Swedish iron work subsidiary which abstained from net investments during a period of 15 years:[85]

Arrows learning modes

No new investments took place except from a minimum number of repairs and replacements. Despite of this, production per hour rose, during this period, with an average of nearly 2% per year - to be compared with a growth of corporate productivity (measured per hour) of nearly 4%. Within the remaining part of the corporation, new investments of a considerable size were undertaken during the same period.

Standardizing — Missions that stay the same

(Lundberg, 1961, p.130-31)

The Arrow reference to the Horndahl effect raises the question of how we should understand learning by doing. Lazonick & Brush (1985) argue that the hypothesis of learning by doing implicitly assumes a pure technical relationship between inputs and outputs which, of necessity, improves with production experience. However, this improvement may partly be social in the sense that work effort and work organisation are determined by "historically specific factors", and in order to investigate on learning by doing the researcher ought to distinguish between "the quality of labor power" and the changes "in the extent to which labor power is actually realized in the form of labor services" (Lazonick & Brush, 1985, pp.55, 57-58), i.e. human skills and the improvement of work organisation.[86]

innovations that persist are standardized
Flicker & ferde?

---

85. My translation from Swedish. The same type of experience has, for instance, been reported from a subunit of a very large Danish manufacturing company, where production, due to an increased focus on the quality of standard operating procedures, grew with 41% during a period of 3 years despite of the absence of new investments (Gjerding, 1990, p.401).

86. In their empirical analysis, Lazonick & Brush (1985) include the distinctions between different types of techniques, quality of input, capacity utilisation, measures of the quality of labour, and various measures to capture the managerial strategy and the degree of workers' control over their

learning by doing in many areas

standard operating procedures ⟹ learning by doing

This line of criticism may be extended by saying that the Arrow model assumes "that learning takes place as a by-product of ordinary production", and even if Arrow (1962) recognises that "society has created institutions, education and research, whose purpose it is to enable learning to take place more rapidly" (ibid., p.172), learning is not percieved as the very dynamics of technical change (as argued in section 6.4). The reason for this is that learning by doing is a type of learning which take place when the operating procedures have become "standard" and the degree of uncertainty negligible - or, in the words of Rosenberg (1982), after learning in research and development has taken place.

Inspired by Rosenberg (1982), we might argue that operating procedures change as the outcome of learning in R&D which represents the search for new ways of doing things. If learning in research and development is localised, then the transition from search to the change of operating procedures, i.e. from R&D learning to learning by doing, takes place at a high speed. And contrary, the less localised search is, the lower the speed of transition. In consequence, we should expect a high degree of localised search to be significantly positively correlated with the growth of productivity. This *hypothesis* is tested in the following. *learning in R+D + localised*

The problemistic search involved in the Nelson & Winter (1974) model provide the firm with the opportunity of developing solutions of its own or imitate solutions present in the external selection environment.[87] It seems reasonable to suggest that R&D spending related to imitative strategies is smaller than R&D spending related to innovative strategies.[88] This assumption, which is in line with the taxonomy of Freeman (1982) presented in the following chapter, is employed by Nelson (1987, pp.43-46) in a recent

spending on imitative vs innovative

job situation. Furthermore, the analysis distinguish between the ethnical background of the workers. Their empirical findings support the technical learning by doing hypothesis, but also the importance of industrial relations. Thus, "the production-relations hypothesis should be given at least as much attention as the learning-by-doing-hypothesis in research into the 'Horndahl effect'. There is much more to the process of labor productivity growth than the technical development of inputs. Social influences on productivity growth must be considered as well" (ibid., p.83). Chapter 8 returns to the issue of human skills and work organisation, however subject to a more limited scope than in Lazonick & Brush (1985).

87. The Nelson & Winter (1974) model was devised in a way that made the subsequent computer simulations comparable to the work of Solow (1957). Production techniques were assumed to be putty-clay, and net profits were assumed to be consumed by gross investments, *irrespective* of the type of strategy employed by the firm. However, although if the latter condition is changed and some firms are allowed to be innovators while others are imitators, as in the Nelson & Winter (1982) modelling of Schumpeterian competition, the basic functioning of the model remains the same. Models of Schumpeterian competition are presented in Nelson & Winter (1982, chapters 12-14), and Nelson (1987, pp.37-43) provides an overview.

88. Chapter 7 returns to this issue.

best practices = imitative
? = innovative

model on Schumpeterian competition. The objective of the model is to validate the assumption that a high research intensity, measured by the ratio of R&D to sales, is positively correlated with relative productivity growth[89], and the conclusion which appears from the working of the model is in accordance with empirical evidence presented by Nelson (1987, pp.47-50). However, irrespective of the strategy chosen, the benefits from technical advance, measured as productivity growth, should be expected to be positively correlated with R&D spending in any manufacturing firm. If this assumption is accepted, for the moment, some empirical evidence on the Danish manufacturing R&D effort during the period of 1974-90, with an emphasis on the subperiod of 1980-90, may provide the means to test the hypothesis stated above.

Table 1 presents the longitudinal distribution of Danish manufacturing R&D according to purpose, divided into three broad groups: Product development, process development, and other purposes. Product-oriented R&D refers to the development of products which are new to the firm, new to the firm but not to the market, and improvement of existig products. Process-oriented R&D refers to the development of new processes and systems, and the improvement of existing processes. Other purposes refer to activities that cannot be classified elsewhere, such as the development of knowledge which cannot be associated directly with product and process development. It has often been argued that product innovation consumes the major part of the resources devoted to R&D, and the Danish evidence presented in table 1 confirms that the Danish manufacturing R&D shows the pattern which is, generally, to be expected (Christensen, J.F., 1992, pp.13-14), i.e. that approximately 3/4 of the innovation effort is devoted to product innovation[90], depending on whether we define innovation from the point of view of an industry or a firm.[91] Table 1 employs the firm as the point of reference for defining an innovation.[92]

---

89. Nelson (1987) employs this model in an inter-industry context, but the positive relationship may be assumed to apply to the level of the firm as well.

90. Measured as the share of the number of important innovations, as in Pavitt (1984), or as the share of manufacturing R&D, as in table 1.

91. If we define product innovation as something brought to use outside the industry which developed it, the number of product innovations becomes smaller than if we employ the firm as the basis for our definition as in table 1.

92. Every second year, the directorate of research (Forskningssekretariatet; from 1988: Forskningsdirektoratet; and from 1990: Forskningsafdelingen) at the Ministry of Education publishes a survey on industrial R&D. Unfortunately, the inclusion of the year of 1989 in table 1 poses some problems, because a new question has been added to the questionnaire from which the figures are compiled. The firms are now asked to give figures on R&D spending on customised product development, and in the 1989 survey approximately 7.6% of the resources devoted to R&D were recorded in this category. In order to make the figures of 1979-87 comparable to 1989,

R + D expenditures in
product & process innovation

Table 1. *The distribution of R&D in the Danish manufacturing sector according to purpose, percentage of total spending, 1979-89*

Purpose	1979	1981	1983	1985	1987	1989
Product development	**77**	**78**	**77**	**77**	**78**	**84**
- new to the market	27	29	24	24	25	27
- new to the firm, but not to the market	18	20	21	19	18	19
- improvement of existing products	32	29	32	34	35	38
Process development	**17**	**16**	**19**	**18**	**16**	**13**
- new processes, systems etc.	9	9	11	9	9	6
- improvement of existing processes	8	7	8	9	7	7
Other purposes	**5**	**5**	**5**	**6**	**6**	**4**
Total	**99**	**99**	**101**	**101**	**100**	**101**

Source: Forskningssekretariatet (1979, 1980, 1982, 1985, 1987), Erhvervslivets forsknings- og udviklingsarbejde *(Industrial R&D), Kobenhavn: Undervisningsministeriet (Ministry of Education); Forskningsdirektoratet (1988),* Erhvervslivets forsknings- og udviklingsarbejde *(Industrial R&D), København: Undervisningsministeriet (Ministry of Education); Forskningsaf-delingen (1990),* Erhvervslivets forsknings- og udviklingsarbejde *(Industrial R&D), København: Undervisningsministeriet (Ministry of Education)*

With the exception of 1989, the time pattern is quite stable, and as shown in table 2, the stability of the distribution of R&D according to purpose is associated with a continous increase in the R&D share of gross domestic product for the period as a whole. However, behind this regularity we find a fluctuating pattern of growth which exhibits small growth rates in the R&D share (below 5%) in 1980, 1984-85 and 1988-90 and high growth rates

---

these 7.6% have been allocated on a 50-50 basis in table 1 to the development of products new to the firm but not to the market, and the improvement of existing products. It is assumed that customised product development is initiated by customers who requires the firm to improve on some product previously supplied to the customer, or to develop a product, the idea of which the customer has obtained by observing what is available at the market. Thus, it is assumed that the share of the development of products new to the market *and* initiated by customers is negligible. This is, of course, a strong assumption, as chapter 7 might indicate.

in 1981-82 and 1986-87.[93] Without considering the industry distribution, which of course implies the heroic assumption that there is no inter-industry differences[94], the impression from table 1 is that the Danish manufacturing sector devotes most of its attention to the improvement of existing product and process technology (36-40%), and the development of new processes and products new to the firm but not to the market (25-32%).[95] This is hardly surprising, since firms tend to focus on the expansion of their business core as long as earnings are satisfactory and prospects promising. The hypothesis presented earlier suggested that in cases where learning in research and development is localised, the transition from R&D learning to learning by doing takes place at a high speed. In this case, R&D results, with some time lag, in an increase of organisational performance, measured in terms of productivity. *Transition from R+D to LBD*

The occurrence of a time lag appears in table 2, where the growth of manufacturing labour productivity and the R&D share in GDP exhibit opposite time patterns. Assuming a linear relationship between productivity growth and the growth of the R&D share, a number of regression analysis on the periods of 1974-90 and 1980-90 showed that any significant relationship appears only in the case of a two-year lag, i.e. $P_t = a + bR_{t-2}$, where P is productivity growth and R is the growth of the R&D share. Figure 31 summarises the regression results, and although the number of observations is rather small in both equations, the satisfactory value of the t-statistics might provide for this problem. The regression results seem to indicate that the transition from R&D learning to learning by doing took place at a higher speed during the eighties than during the whole period of 1974-90, as reflected in the higher values of the regression coefficient and the rate of increase of the regression curve (b). At the same time, it appears that the growth of labour productivity becomes more sensitive to the growth of the R&D share. This result is surprising since the distribution of R&D spending among familiar and unfamiliar activities is quite stable, cf. table 1, and the stability of the R&D distribution indicates that the regression results should be interpreted with some care, for a number of reasons.

---

93. The rate of manufacturing profit, defined as the ratio between gross operating surplus and gross domestic product, increased in the first half of the period, inducing manufacturing investments to boom in 1983-86 (Gjerding, 1991). The investment boom was, in general, substantiated by an increase in the growth of manufacturing output, and during such a period of high levels of economic activities, the R&D share in GDP may be expected to decline, and *vice versa* in the case of low levels of economic activity. However, I am not in a position to validate this suggestion.

94. Which there is, of course. Chapter 7 returns to this issue.

95. The development of new products (24-29%) consumes, in general, one quarter of the R&D expenditure.

Table 2. *The growth of labour productivity measured as GDP per hour at 1980 prices, and the share of GDP devoted to R&D in current prices, Danish manufacturing, percentage changes, 1980-90*

Year	Growth in productivity	Growth in R&D share*	R&D share*
1980	4.96	0.00	2.37
1981	2.34	12.24	2.66
1982	-0.47	9.02	2,90
1983	6.63	5.52	3.06
1984	0.24	-1.96	3.00
1985	-0.08	2.00	3.06
1986	-5.10	9.15	3.34
1987	0.05	11.38	3.72
1988	3.46	1.88	3.79
1989	5.16	2.37	3.88
1990	1.19	1.29	3.93

Source: Calculated from the Danish national accounts and OECD (1992), Business Enterprise Expenditure on R&D in OECD Countries, *Paris: OECD*
* *R&D figures for even years (1980, 1982 etc.) are based on OECD estimates*

First, the regression analysis rests on the dual assumption that (1) since learning by doing may be measured by the growth of labour productivity, (2) the transition from R&D learning to learning by doing may be measured by the relationship between the growth of labour productivity and the growth of the share of R&D in manufacturing activity.[96] However, the relationship is far from unidirectional, since changes in the R&D effort may be directed towards activities that do not translate into learning by doing, and since learning by doing may occur in the absence of R&D as the Horndal experience indicates. In consequence, we are dealing with proxies that measure proxies.

---

96. It seems reasonable to adopt the growth rates rather than the levels to capture this transition.

Figure 31. *Regression analysis of the linear relationship between the growth of labour productivity measured as GDP per hour (P) and R&D share of GDP (R), regression on $P_t = a + bR_{t-2}$, Danish manufacturing 1974-90 and 1980-90*

Period	1980-90	1974-90
a	-1.92	0.64
b	86	35
R^2	75	25
t-value	452	209
prob	0	6
D.W.	143	125
D.F.	7	13

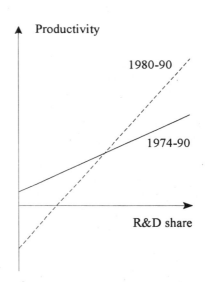

Source: Calculated from table 2 with an extension of the time series to cover the period of 1974-80

Second, autocorrelation may exist bewteen the growth of labour productivity and R&D share, since both must be assumed to be positively correlated with a number of background variables such as the development of turnover and profits. However, the satisfactory value of the Durbin-Watson test might provide for this problem. Furthermore, it can be observed from table 2 that the R&D share is increasing throughout the period irrespective of the business advances and repercussions of the Danish manufacturing sector.[97] This may indicate that the development of the R&D share is influenced by corporate strategy to a larger extent than the development of productivity and thus may be described as a "genuine" regressor.

Third, it should be remembered that we are dealing with a simple linear regression where some indirect effects are bound to influence the value of the regression coefficient. As chapters 7-8 indicate, the R&D spending may be associated with barriers to the

---

97. The Danish manufacturing sector experienced a rapid increase in profits during the first part of the 1980s followed by a decline during the mid-80s and a renewed increase during the last part of the 1980s. Concomitantly, the economic activity boomed during the mid-80s, followed by a severe set-back (Gjerding et al., 1990; Gjerding, 1991).

initiation → implementation
feed back

transition from initiation to implementation, which occur because the transition from initiation to implementation requires the operation of feedback mechanisms, as argued in chapter 2. One such important feedback mechanism is the outcome of *learning by using*, as proposed by Rosenberg (1982) who, with respect to a given product, distinguishes between

LBU

> gains that are internal to the production process (doing) and gains that are generated as a result of subsequent use of that product (using). For in an economy with complex new technologies, there are essential aspects of learning that are a function not of the experience involved in producing the product, but of its *utilization* by the final user. This is particularly important in the case of capital goods. Thus, learning by using refers to a very different locus of learning than does learning by doing. There are various reasons why this should be so. Perhaps in most general terms, the performance characteristics of durable capital goods cannot be understood until after prolonged experience with it. [...] The learning-by-using experience creates two very different kinds of useful knowledge that, borrowing from a well-established terminology, we may designate as *embodied* or *disembodied*.

embodied & disembodied (Rosenberg, 1982, pp. 122; 123)

While the embodied type of knowledge refers to knowledge which leads to an improvement in product design through feedback to the stage of development, the disembodied type of knowledge refers to knowledge of "certain alterations in use that require no (or only trivial) modifications in hardware design" (ibid., p. 124). Thus, in the case of learning by using, the embodied type of knowledge feed back on the process of search while the disembodied type of knowledge feed back on the operating procedures. Consequently, only the embodied type of knowledge influences the *transition* from search to operating procedures directly. However, this generalisation is only true in the short and medium run. In the long run, the disembodied type of knowledge may influence the type of search which will take place since it exerts a direct influence on the operating procedures which define the agenda for that search.   *Flexibility / Stability approach*

The implications for the flexibility-stability approach implied by the basic evolutionary approach described in the present section may be summarised in the following way: What has been discussed so far points to the interplay between operating procedures, investment rules and search procedures as dependent on the type of learning processes involved. Learning by doing results in a gradual change of operating procedures and thus a gradual change in the agenda of localised search. Learning by using influences the medium and long term change of operating procedures through (1) its influence on the processes of search in the case of embodied knowledge, and (2) its influence on operating procedures

embodied & search

144

transition of search to operating procedures

in the case of disembodied knowledge.[98] In consequence, learning by using may to a larger extent than learning by doing lead to types of search processes which are less localised. Similarly, learning by doing is to a larger extent than learning by using associated with organisational changes that imply adjustment learning, and *vice versa* in the case of turnover learning. While learning by using entails a feedback loop generated by the interplay between the internal and the external selection mechanisms, this type of feedback loops occurs less frequently in the case of learning by doing.[99]

look at adjustment & turnover *obstacles & localized*

6.4. Learning as an interactive process

The transition from R&D learning to learning by doing may be impeded for a number of reasons. Following the neo-behavioural approach outlined in section 6.2, such obstacles may occur as the result of pathologies in experiential learning. Following the basic evolutionary approach outlined in the previous section, obstacles relate to the extent that search become increasingly less localised, as far as the short and medium term is considered. In both cases, organisational action is impeded because the organisational context involves rigidity, since organisational learning results in organisational action only to the extent that the theory of action implies that organisational action should be taken.[100] Since it is easier to modify decision rules than to change the theory of action which determines the decision rules, there is a tendency towards *gradual* change in the rules that define how responses are assembled and a tendency towards *discrete* change in the rules which define how stimuli are interpreted. This difference reflects that the internal selection mechanism operates with some rigidity, and that some parts of the organisational repertoire has to be removed if organisational action is to become effective. *removing parts*

However, this point of view does not imply that the rigidity of the internal selection mechanism is, in itself, an obstacle to organisational learning. The flexibility-stability

Learning by scaling

---

98. Learning by scaling (Sahal, 1985) may be interpreted as an example of learning by using. Learning by scaling refers to the kind of learning involved in the fight against the constraints imposed on the innovation process when the size of the innovation is changed, and Sahal (1985) points to three major types of innovations: *Structural* innovation that refers to the improvement of the compatibility of the different parts of the innovation; *material* innovation that refers to the change and economisation of the material employed in construction; and *systems* innovation that refer to the complexity of the innovation in question. These types of learning refer, in the context of learning by using, to the effect of embodied knowledge.

99. In the Arrow sense. If one employs the more broad notion suggested by Lazonick & Brush (1985), learning by doing is interpreted in terms of industrial relations which imply the interplay between internal and external selection mechanisms that are more broadly defined than in the Nelson & Winter theory, since the latter does not include socio-political aspects.

100. As evident from chapters 2-4 and demonstrated in chapter 5.

sociopolitical aspects of learning

145

approach implies that during the stage of implementation, organisational activities have to become routinised in order to secure a productive outcome. The way in which organisational activities are coordinated through the formal and informal organisation structure imposes constraints on organisational action, and these constraints are necessary if concerted action is to become effective and an agenda for future search is to be defined. Thus, since innovation does only occur through some form of organisational learning, the innovation process may, itself, be described in terms of rigidities that shape the directions for future organisational learning. *innovation in terms of rigidities*

These rigidities, identified as the organisational set-up, present the organisation member with a socially ordered context that makes the actions of other organisation members understandable and predictable, however to varying degree as indicated by the previous chapters. This implies that organisational learning is an interactive process. From this point of view it may be argued that learning is "a socially embedded process which cannot be understood without taking into consideration its institutional and cultural context" (Lundvall, 1992a, p.1). Accordingly, innovation, as the outcome of organisational learning, "is shaped by institutions and institutional change" (Johnson, 1992, p.24). As pointed out by Johnson (1992), the existence of institutions does not imply that innovation are impeded, but rather that innovation takes place in a social context characterised by regularities of behaviour.[101] Institutions may broadly be defined as

Definition of institutions ✱

sets of habits, routines, rules, norms and laws, which regulate the relations between people and shape human interaction. By reducing uncertainty and, thus, the amount of information needed for individual and collective action, institutions are fundamental building blocks in all societies.

– organizational set up
– socially ordered
(Johnson, 1992, p.26)

Johnson (1992) proposes an institutional model of the causality involved in the relationship between learning and innovation, cf. figure 32 which describes the learning-innovation relationship as a process of feeding and feedback in terms of a number of learning concepts.

✱ institutional model of causality

101. Describing the contrary point of view, Johnson (1992) points to the institutional drag hypothesis which implies that technical change is retarded by institutional inertia, as emphasised in for instance the late 1970s debate on *institutional sclerosis* (Johnson, 1981). While the institutional sclerosis hypothesis was promoted by some parts of the OECD analytical apparatus, later work undertaken by other parts of the OECD apparatus pointed to the existence of an institutional-technological mismatch which disconnected economic growth from technical progress in the OECD area (Johnson, 1992, p.24). The latter hypothesis appeared in the work by Sundqvist et al. (1988), and in a Danish context this hypothesis was promoted by Gjerding et al. (1988) and later investigated in Gjerding et al. (1990). Chapters 7-8 return to this issue.

institutional / technical mismatch

Figure 32. *The interplay of learning and innovation from an institutional point of view*

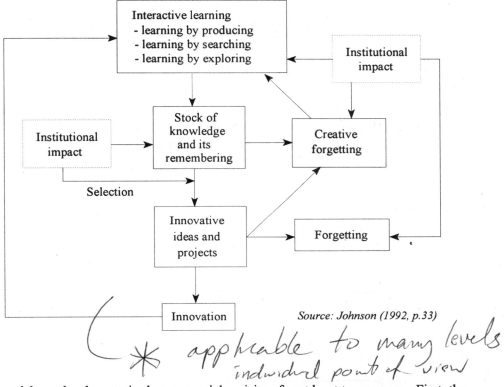

Source: Johnson (1992, p.33)

The model may be characterised as *general* theorising, for at least two reasons: First, the model refers to basic processes of human cognition and thus makes the process of learning understandable from the point of view of the individual capacity for learning. Second, the generality of the model makes it applicable to many levels of aggregation, i.e. the level of intraorganisation, interorganisation, economy and intereconomy. In consequence, the model describes the relationship between institutions and learning in a way that identify some basic regularities which may be found at all levels of the economic system. However, in the present context, further discussions of the model takes place at the level of the organisation.

The causality of the learning-innovation link is determined by the way in which learning affects the accumulation and reproduction of the stock of knowledge, and the flow of innovative ideas and projects stemming from that stock.[102] Additions to the stock

---

102. This way of describing the Johnson model is inspired by Levinthal (1992) who distinguishes between stock and flow variables in the analysis of organisational survival and talks about a stock

information processing model?

of knowledge are continuously provided by learning, but substractions do appear as the deterioration of knowledge which occurs when knowledge is not "actively remembered", i.e. forgotten. Forgetting may be more or less instrumental to the innovative process in that some types of forgetting may be related to de-learning and other types to simply forgetting as such. In the case of *de-learning*, old habits of thought and routines vanish from the organisational repertoire because the organisational learning implies the addition of new parts to the repertoire which become effective only as some old parts are substracted. This is a process of "creative forgetting", by implicit analogy to the Schumpeterian concept of creative destruction.

Creative forgetting ✱

> Forgetting is, thus, an essential and integrated part of learning, even if it is not always easy to seperate *ex ante* between "creative forgetting" and "just forgetting".
>
> *↑ interaction   technical*   (Johnson, 1992, p.29)

The learning concepts that appear in figure 32 reflect the point of view that learning may be ranked on a scale of increasing human interaction. Types of learning related to technical change are, typically, located at the high-frequency part of the scale since technical change often requires "sequences of exchanges of messages between different people in different departments and at different levels, within firms and between firms" (ibid., p.31). At the most high-frequency part of the scale one may find *learning by searching*, which refers to types of profit-oriented learning explicitly intended to stimulate and organise the increase of the stock of knowledge, and *learning by exploring*, which refers to search activities that take place in, generally, non-profit organisations such as universities. In the middle range, one may find *learning by producing*, which refers to processes of learning related to the normal production activities of the organisation. These processes may be described in terms of learning by doing, learning by using and learning by interacting (i.e. the concept of learning by producing is a synthesis). *Learning by interacting* was, originally, proposed by Lundvall (1985) as an interorganisational concept[103] that described the outcome of feedback processes between producers and

high

middle

---

of organisational capital (Levinthal, 1991), aspiration levels as stock phenomena (since they reflect both past and current outcome) and performance as a flow phenomenon that feed into the aspiration level(s).

103. The concept provides a tool to describe why markets become organised for reasons other than those invoked by the market failure approach (Williamson, 1975). As employed by Lundvall (1988), the concept of interactive learning is an institutional concept which describes the organisation of markets within a national framework. Examples of the importance of institutional boundaries to learning presented by the specific characteristics of the national setting appear throughout the various contributions in Lundvall (1992), and Kogut (1991) and Johnson &

*institutional concept
= interactive learning*

professional users which in some instances are characterised by information assymmetries. However, from the point of view of the *general* theorising implied by the model depicted in figure 32, learning by interacting may be interpreted as an intraorganisational concept as well, since inside organisations, as between organisations, innovation is contingent upon the establishment of channels for communication, coordination and control, which may be characterised by the presence or absence of information assymmetries. *LBI intraorganisatn*   *LBS deliberate*

Learning by searching reflects that the organisation deliberately organise for the accumulation of the stock of knowledge. In most firms, learning by searching is undertaken by a special part of the organisation, e.g. by key personnel, or by a R&D department if the organisation is devoted with the amount of resources necessary to specialise some part of it in deliberate search. Learning by searching may be described more broadly by saying that learning, like production, becomes specialised as a consequence of the division of labour, in a deliberate attempt to learn to learn (Stiglitz, 1987). However, following Argyris & Schön (1978), learning to learn implies more than the specialisation of organisational resources devoted to learning by searching.

According to Argyris & Schön (1978), one may distinguish between single-loop, double-loop and deutero-learning.[104] These concepts are more broad and comprehensive than the other learning concepts described so far in the present chapter. While *single-loop* learning refers to the case in which performance gaps are remedied "by changing organizational strategies and assumptions within a constant framework of norms for performance" (Argyris & Schön, 1978, pp.20-21), *double-loop* learning refers to the case in which the set of norms have to be altered in order to overcome the performance gap, either by changing the priorities among norms or by changing the set (ibid., p.24). To the extent that the organisation learns how to master single- and double-loop learning, we may speak of *deutero*-learning as a concept which coins the process of learning to learn (Bateson, 1942).

This process is illustrated in figure 33. If we imagine a case in which an organisation member is confronted with a performance gap at three successive points of time, rational problem-solving in the behavioural sense will gradually lead to a faster and more effective response as time goes by, since those problem-solving activities that proved valuable in the past tend to become reinforced. This is depicted in the upper part of figure 33 which shows the learning curves that emante as the aspiration level of the organisation member

---

Lundvall (1992) provide interesting treatments of the aspect of institutional and organisational learning at the macro level.

104. Reference to double-loop learning was made previously in section 5.4.

Learning becomes specialized like production

149

is met at the three successive points in time [t;t+2]. The lower part of figure 33 depicts the ensuing deutero-learning curve that can be derived from the three successive learning curves. The deutero-learning curve is the mirror reflection of the three learning curves in the sense that while the learning curves tend to grow at an increasing rate of problem-solving, the rate of increase declines in the case of the deutero-learning curve. This implies that the organisation member in question becomes increasingly more effective in his problem-solving activities. At the organisational level, the various pathologies described in terms of experiential learning may occur with an ensuing tendency to slow down the rate of problem-solving and thus lower the rate of increase of the deutero-learning curve. In the case of the transition from R&D learning to learning by doing discussed in section 6.3, the declining rate of increase of the deutero-learning curve will be reflected in a declining rate of productivity growth. However, in the case of learning by using, the rate of increase of the deutero-learning curve will tend to increase to the extent that the embodied type of knowledge feed back on the process of search, and in the long term the effect of the disembodied type of knowledge may provide a stimuli for the rate of increase of the deutero-learning curve as the agenda of search becomes gradually redefined - provided, of course, that the accumulation of the stock of knowledge is not impeded by the pathologies of experiential learning. It is to this issue that the last section of the present chapter devotes its attention.

Figure 33. *Deutero-learning curve derived from three successive learning curves*

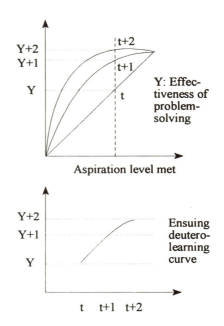

Source: Adapted from Bateson (1942, p.168)

6.5. Intersections of neo-behavioural and evolutionary concepts of learning

The concluding discussion of the present chapter takes, as its point of departure, the assumption that it is possible to interpret the evolutionary concepts of learning discussed in sections 6.3-6.4 in terms of the concept of experiential learning presented in section 6.2, with the exception that the pathologies described earlier does only occur to a limited extent in the cases of learning by doing and learning by interacting. The exceptions are

150

depicted in figure 34 as dotted cells, while the bracketed numbers refer to the cases where interpretation may take place.

Figure 34. *Elaborating the concept of experiential learning*

Learning	Audience	Role-constrained	Superstitious	Under ambiguity
By doing		(1)		
By using	(2)	(3)	(4)	(5)
By interacting	(6)	(6)	(7)	(6)
By searching	(8)	(8)	(8)	(8)
By exploring	(8)	(8)	(8)	(8)

Regarding the case of *learning by doing*, audience learning, which refers to the decoupling of individual and organisational action, is not easy to concieve. Learning by doing, being instrumental to the growth of labour productivity, will inevitably increase the level of organisational performance and cannot be meaningfully described without reference to its effects on operating procedures. Since learning by doing refers to the gradually increasing proficiency in operating procedures, superstitious learning and learning under ambiguity, which refer to the relationship between the individual, the environment and the individual beliefs, are also not easy to concieve, at least in the case of learning by doing in the technical Arrow sense. However, *(1)* learning by doing may be role-constrained in the sense that during the course of daily operations, the organisation member may confine himself to the procedures defined by his organisational role, although his past experience leads him to believe that things might be done more effectively if the boundaries of his organisational role were removed.

In the case of *learning by using*, four cases may occur. Although learning by using (in the embodied sense) "permit subsequent improvements", as Rosenberg (1982, p.123) puts it, and is "associated with a high degree of systemic complexity" (ibid., p.135), it may, nevertheless, take the form of *(2)* audience learning since experience accumulated through learning by using at the level of the organisation member may not result in a growth of the stock of knowledge. For instance, code scheme barriers, insufficient communication channels or problems of information overload may prevent the accumulated knowledge of the organisation member to translate into organisational action. Similarly, *(3)* role-constrained learning may occur since, for instance, the organisation member may refrain

selection

preventing knowledge from → org. action

role constrained in the case of teachers & architects fh instrs

from feeding back his knowledge acquired through using-experience if the role of the organisation member does not imply that such action should be taken. This situation would appear in a case where some operator's experience with a CNC lathe is never fed back to the CNC producer, because the operator does not interact with the production manager who has the customer-seller relationship with the CNC producer. *(4)* Superstitious learning may occur in the later parts of the development feedback process, for instance because the CNC producer interprets the needs of the CNC user erroneously. Finally, *(5)* the systemic nature of learning by using involves a complex set of different organisational agents and thus is likely to appear as learning under ambiguity, at least until some state of dominant design has been established.

In the case of *learning by interacting* in the Lundvall (1985) sense, audience and role-constrained learning cannot, *per definitionem*, occur since learning by interacting refers to feedback processess above the level of the individual. However, *(6)* to the extent that learning by interacting is influenced by information assymmetries or result in unsatisfactory innovations, as described by Lundvall (1985), the perverse effects of learning by interacting may be explained by reference to audience and role-constrained learning, and learning under ambiguity. The most direct relationship between learning by interacting and experiential learning appears in the case of *(7)* superstitious learning where organisational action has but a little or even unanticipated effects on the environmental response. This is, for instance, the case of hyper-centralisation described by Lundvall (1985, p.41-43) where a number of computer users developed their own data processing units because they were unable to use the systems and programs provided by a data processing and office technology producer who developed systems and programs that "were more centralized in their design than both technical opportunities and user needs could infer" (ibid., p.43). A similar situation appeared in the Danish dairy industry where

information assymetries

dairy processing plants designed by the producers of equipment and systems were more capital intensive, more inflexible, and *more highly automated* than what corresponded to cost-effective solutions and the needs of the users. (Lundvall, 1985, p.33)

↑ dairy

perverse effects of computers

Learning by searching and learning by exploring refer to the types of search activities deliberately organised in, respectively, profit and non-profit oriented organisations in order to increase the stock of knowledge. As described in the previous section, these types of learning are found at the high-frequency end of the scale of human interaction. Since both concepts reflect individual, intra- and interorganisational phenomena, *(8)* audience, role-constrained and superstitious learning and learning under ambiguity may be involved.

How relate to K-12 systems set for districts & accountability rather than teaching

Chapter 7

Paradigms and Activities in the Innovation Process

7.1. Innovation as a process of feedback

innovative_us. imitative

The purpose of the present chapter is to relax some of the strong assumptions about the innovation process which have been present, partially explicit, in the preceding chapters. *First*, it has been assumed that it was possible to describe the process of learning generally irrespective of whether we are dealing with an innovative or imitative endeavour. *Second*, the discussions have not taken into consideration the elements of sectoral patterns of innovation and interorganisational interaction along sectoral patterns. *Third*, the degree of uncertainty entailed in innovation has only been treated in terms of more or less localised search, e.g. the idea that the less localised search is the more uncertain is the outcome of innovation. *Fourth*, the emphasis of the preceding discussions have been on process and organisational innovation. However, the flexibility-stability approach is applicable to product innovation as well, *and* to the *interplay* between technical (process and product) and organisational innovation, as appears from chapters 2 and 8. While chapter 6 aligned theories of learning processes within the field of organisation theory and innovation economics, the emphasis of the present chapter is on some models of the process of technical innovation which have appeared in the field of innovation economics. Having presented these models, chapter 8 returns to the alignment of organisation theory and innovation economics.

Following Saren (1984), innovation process models may be classified according to their breakdown of the process in its constituent parts and the ensuing focus of attention on one stage or factor. Figure 35 summarises the classification offered by his review.

Department-stage models assume that the innovation process takes place in a series of stages, each of which is associated with an intrafirm department delineated by clear and stable boundaries. The innovation may be described as an idea that travels through R&D to design, further on to engineering and production, increasingly becoming manifested in physical form and, finally, approaching the market through the channels developed by marketing and sales. *innovation in stage*

Activity-stage models assume that the innovation process can be broken down into a sequence of activities that may take the form of, for instance, idea generation, problem-solving and implementation, or planning, development and evaluation. The various activities may take place in specific departments, or they may be dissociated from specific

activities

flexibility/stability approach applicable to process, organisational, & product innovation

153

departments. In either case, the analytical emphasis is on activity rather than **departmental** allocation.

Figure 35. *Five models of the innovation process*

Type of model	Focus of attention of the model
Department-stage	Stages associated with intrafirm departments
Activity-stage	Stages associated with innovation activities
Decision-stage	Stages of information-gathering in relation to decision points
Conversion-process	R&D phenomena; the process of innovation as a "black box"
Response	S-R mechanisms associated with organisation members

Source: Saren (1984)

Decision-stage models describe the innovation process as a series of information-gathering activities which leads to decisions that may be classified as key decisions related to action points. As a main characteristic, the number of decision points, which can be identified by the decision makers, determines the number of stages of information-gathering.

Conversion-process models view the innovation process as a conversion of inputs to outputs, where the choice among inputs is determined by the overall strategy of the firm. The focus of attention is on factors such as the rate and direction of technical change, the related factor costs, and the intrafirm allocation of scarce resources to various types of activities (notably the R&D budget) according to departmental and/or activity-related stages of innovation. The conversion-process models comprise many of the features found in department- and activity-stage models, but abstain from identifying the constituent parts of the innovation process.

Finally, *response* models describe the innovation process as a reaction to internal or external stimuli which force one or more organisation member to concieve a possible solution to some kind of problem. The stimuli-response mechanism involved takes place as the conception of an innovative idea that occurs as a response to some kind of stimulus and is percieved through search for and evaluation of the solution and its implications. The concieved idea is championed through various types of proposals and ensuing

Models of the innovative process

decisions to adopt or reject the innovation, which may lead to an organisational response. In general, response models emphasise the initial, individual-oriented parts of the decision process to some neglect of the later, organisation-oriented, parts of the stimuli-response mechanism involved.

Conversion-process models abstract, in general, from the constituent parts of the innovation process. Department-stage models are, like the conversion-process models, quite useful in order to identify the formal task environment of a particular innovative activity, but unable to capture the decision making nature of the formal task environment. Activity-stage models describe the *kind* of activities going on, but the innovation process appears as being directed, to an extreme degree, by a well-defined agenda since they give "no indication of the *alternative paths* available" (Saren, 1984, p.17). The aspect of alternative paths is present in the decision-stage models where decision points are identified and the choice is something more than a binary choice of do or don't, but they do not include the description of ongoing activities so prominent in the activity-stage models. The reponse models identify the important mechanism of the occurrence of performance gaps and the subsequent change of aspiration levels, but suffer from a tendency to overemphasise the initial, individual-oriented, parts of the innovation process at the cost of the analysis of the later, organisation-oriented, parts of the innovation process which can be associated with the transition from the stage of individual or subunit proposal to the stage of organisational adoption/rejection.

However, this type of criticism should not be interpreted as if the models reviewed by Saren (1984) do not yield important and interesting insights. On the contrary, they do, and the empirical results referred to in chapter 8 are, mostly, derived within models that can be described in terms of figure 35. Any model is, of course, an abstraction from the real world and can only yield results in accordance with the analytical limitations of the model. One cannot but be in agreement with Saren (1984) when he argues:

model as abstraction

> Of course, the accuracy of the model used depends largely on what it is being used for. (...) So what I am *not* going to conclude in this paper is that one approach is better than another, since this depends on its purpose. [...] A general model is required which accurately segments and describes what occurs in the firm during the innovation process.
> (Saren, 1984, pp.23; 24)

As has been indicated by the previous chapters 2-6, a general model of the innovation process must emphasise the behavioural regularities of organisational action and stress the importance of feedback mechanisms in interactive learning. The unifying characteristic of the models reviewed by Saren (1984) is the *sequential* nature of analytical reasoning, and they may be treated as representatives of what Kline & Rosenberg (1986) have termed

feedback + regularities

155

the linear model which describes the innovation process as a progression from (re)search to marketing via development and production. Kline & Rosenberg (1986) argue that the linear model underestimates or even neglects the process of learning through feedback as an integrated part of the innovation process, and present, as an alternative, the *chain-linked* model which comprises a central core of progressive stages interlinked by a complicated structure of feedback processes as learning takes place in relation to the evolution of a dominant design.

Figure 36 presents a simplified version of the chain-linked model. The downward sloping structure of five boxes represents the main chain of activities which runs from potential to actual market needs through design and production and is linked by feedback loops. At various points the constituent parts of the main chain are interlinked with the existing stock of knowledge, and if the existing stock of knowledge is sufficient to solve the problem at hand research will not be undertaken. To the extent that the existing stock of knowledge proves to be insufficient, research is undertaken; however, it may be far

Figure 36. *The chain-linked model*

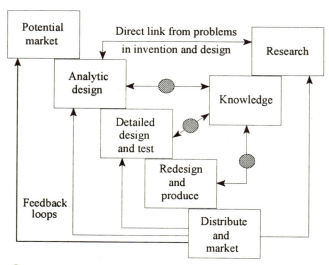

Links through knowledge to research and return paths

— Particularly important feedback

Source: Kline & Rosenberg (1986)

156

existing stock 1st
— innovator as problem solving

more difficult to obtain a solution from research than from utilising the existing stock of knowledge (hence the dotted circle). The main difference between the linear model and the chain-linked model relates primarily to how the learning process is described. *First*, the two models differ regarding the role of learning through feedback. While the linear model, largely, ignores the learning process associated with normal production activities, the chain-linked model emphasises learning by feedback from design, testing, production and marketing. *Second*, the two models differ regarding the role of science. While the linear model, in general, percieves technical innovation as applied science and thus has a tendency to adopt a science-push perspective on innovation, the chain-linked model emphasises the role of research primarily as a solver of problems that enter the stage of decision when the available stock of knowledge falls short of solving the problem. This does not imply that science has no part in the play staged by the chain-linked model, since "the use of accumulated knowledge called modern science is essential to modern innovation; it is a necessary and often crucial part of technical innovation, but it is not usually the initiating step" (Kline & Rosenberg, 1986, p.291). In consequence, the chain-linked model is designed to avoid the distinction between a technology-push and a demand-pull causality of innovation which have pervaded a number of innovation studies.[105] → what about in K-12?

Due to the avoidance of the technology-push/demand-pull dichotomy and the emphasis on feedback processes as a propeller for technical innovation, the chain-linked model is endowed with a large amount of analytical use-value. The most important merit of the model seems to be that it singles out the various steps in development which must be undertaken before effective market supply can occur. However, the chain-linked model, suggestive as it is, suffers from some difficulties.

First, even though the generality of the model is one of its great strengths, it is also one of its main weaknesses. One the one hand, the model stresses the main link of activities and singles out important interfaces between main activities and (re)search. These interfaces are quite useful as focuses of attention in the analysis of technical innovation. On the other hand, the chain-linked model may assume various appearances according to the nature of the cases on which the model is applied. For instance, in some cases the innovative endeavour is biased towards the detailed design stage, in order cases towards the redesign and production stage. Thus, while serving as an analytical guide to the

---

105. Mowery & Rosenberg (1979) present a critical discussion of the influence of market demand on innovation. Coombs, Saviotti & Walsh (1987, pp.94-103) present an overview of the demand-pull technology-push controversy.

for info on
demand pull - tech push

157

innovation process, the specific use of the chain-linked model requires that we identify the dominance of some types of main activities. Section 7.2 returns to this issue.

Second, even though the chain-linked model offers an alternative to the traditional linear description of the innovation process, its analytical properties are still confined within the field of innovation economics. Consequently, the model does not emphasise the nature of behavioural regularities which underlies the process of technical innovation. This is, of course, not a critique since this is exactly what the model was designed for; it is merely stated in order to point out that the chain-linked model encounters some analytical deficiencies when applied to cases where organisational innovation, or the interplay between organisational and technical innovation, are important.

In consequence, if one should strive for the general model warranted by Saren (1984), the chain-linked model would be extremely useful as a starting point, but would have to be elaborated in terms of the focus of innovative activity, the behavioural regularities underlying organisational action, and the specific learning processes described throughout chapters 3-6. From the point of view that Saren (1984) might take, it seems reasonable to suggest that a more satisfactory model might be obtained by combining the chain-linked model with the identification of (1) formal task environments, based on department-stage and conversion-process models, and (2) the occurrence of stimuli, activities and decision points, based on the response, activity-stage and decision-stage models. However, in order to capture the nature of organisational learning described in the preceding chapters 3-6, a combined model would have to be elaborated in terms of the existence of routines, routines for changing routines, and intraorganisational conflict on routines.

Christensen, J.F. (1992, p.97) have made the distinction between *simple ideal-type models*, which represent a "functional phase dynamics", and *complex real-type models*. While the nature of linear models is functional phase dynamic, the chain-linked model is an example of the complex real-type models. However inclined to argue in favour of the complex real-type models, Christensen, J.F. (1992) advances the argument that the chain-linked model only to a very limited extent avoids the technology-push/demand-pull dichotomy, and that the model, furthermore, is not designed to define and analyse the duration of the innovation process.[106] *First*, even though the chain-linked model emphasises feedback mechanisms between the parts of the main link and between the main link and the knowledge/research components, only two types of activities within the model seem to initiate innovation: The particularly important feedback, cf. figure 36, "which implies that market assessment of existing or new products leads to concepts for new products" (ibid., p.117); and the direct link between research and problems in

---

106. The following quotations from Christensen, J.F. (1992) are translated from Danish by me.

invention and design. In consequence, "the chain-linked model retains the classic push-pull dualism, even if the model does recognise the occurrence of innovation initiated by both research and markets and thus transcends some stereotypical monolithic conceptions" (ibid.). *Second*, the duration of the innovation process is confined to the second, third and fourth part of the main link in figure 36, with an emphasis on the generation of analytic design, and Kline & Rosenberg (1986) do not give an account of how the first and fifth part might initiate innovation (Christensen, J.F., 1992, p.114). This raises a number of questions:

> To what extent must the individual stages be seperated in terms of time and organisational affiliation in order to be discharged as part of the same process? To what extent does the analysis focus on past events, i.e. adopt an "archeological" point of view in order to detect events upstream or prior to "analytic design"? Furthermore, to what extent does the analysis pursue future events downstream after the initial market introduction?
>
> (Christensen, J.F., 1992, pp.114-15)

In order to analyse these questions, Christensen, J.F. (1992, pp.115-16) suggests the image of an extended process of innovation, cf. figure 37. The chain-linked model comprises what is termed the central process of innovation in figure 37, i.e. that part of the innovation process which leads from an innovation proposal to the resulting market introduction. However, prior to the central process of innovation is the pre-innovation process in which the organisation "generates ideas for new products, makes priorities among projects, and identifies the technology necessary for completing the future projects" (ibid., p.115). Furthermore, the central process of innovation is succeeded by the post-innovation process in which the ensuing product is modified according to changes in market needs and competitive conditions. Following the argument of Adler (1989), Christensen, J.F. (1992)

Figure 37. *An extended process of innovation*

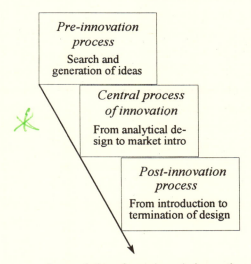

Accumulation of experience in innovation

Source: Christensen, J.F. (1992, p.116)

159

argues that the post-innovation process is conducive to organisational learning in the sense that the experience obtained through the post-innovation process enters the stock of knowledge.

As argued previously, the chain-linked model does not emphasise the behavioural regularities of organisational action which underlies technical innovation. From chapters 2-6 it follows that these behavioural regularities determine the way in which learning takes place by defining the focus of attention and the agenda of search entailed in organisational learning. Inspired by Van de Ven et al. (1989), Christensen, J.F. (1992, pp.117-31) presents two complex real-type models, the *accumulation* model and the *envelope* model, summarised in figure 38 which compares the two models to the classic flexibility-stability approach presented in figure 8.

Figure 38. *Comparing the accumulation and envelope models in terms of the classic flexibility-stability approach presented in figure 8*

Hage & Aiken (1970)	Zaltman et al. (1973)	Accumulation model	Envelope model
Evaluation	Knowledge-awareness Formation of attitudes	Initiation	Initiation
Initiation	Decision		
Implementation	Initial implementation	Accumulation	Linear stage Ramification Convergence
Routinisation	Continued-sustained implementation		Implementation

Source: Figure 8 and Christensen, J.F. (1992, pp.117-31)

The accumulation model focusses *per definitionem* on the stage of accumulation where the innovation project "takes shape through the accumulation of resources (capital, competencies etc.) from internal and external sources" (Christensen, J.F., 1992, p.118). The accumulation model describes the basic sociology of the innovation process, i.e. the struggle between different parts of the organisation for command of internal and/or external resources. Following upon the initiation of an innovation project, the relative strength of command of resources shifts from those in charge of ressource allocation to those in charge of the *use* of resources allocated to the innovation project in question.

160   *accumulation model*
    = sociology of innovation
        power & control

From the point of view of the latter group of organisation members, those in charge of ressource allocation may be defined as external agents and those directly involved in the innovation project as internal agents. In an intraorganisational context, the external agents may be top management, while the internal agents may be the R&D department or some sort of task force.

The way in which the relative strength of command of resources changes depends on the type of innovation project undertaken:

Incremental & power

> In the case of typical incremental innovation, the change of power may often take place in a frictionless manner, possibly in a formal fashion at the beginning of the project, however subjected to the opportunity of external interference. In the case of more demanding and uncertain projects, misunderstandings and setbacks are part of the process and may easily lead to power struggles between internal and external agents regarding the future course of the project (and, perhaps, if the project should be continued at all). Consequently, the process of innovation may take the form of recurrent power struggles where the decisive influence on the process may fluctuate between the internal and external agents; and the presence of several internal and external agents may act as a further complication.

(Christensen, J.F., 1992, pp.119-20)

Schools

The recurrent struggles reflect, in the vocabulary of chapters 3-6, the set of conflicting intraorganisational rationalities, i.e. the set of various goals and aspiration levels based on asymmetries in information and knowledge. For instance, internal agents may form "hyper-optimistic" plans in order to gain external support and get away with it because of information asymmetries. This may happen even in the absence of opportunistic behaviour, since hyper-optimism may reflect a genuine entusiasm that impedes the capability of the internal agents to properly evaluate and assess the implications and state of the innovation project at hand. Furthermore, the external agents may apply myopic evaluation criteria such as the overstepping of deadlines and budgets and judge such phenomena as evidencing that the project is loosing track. To some extent, the occurrence of these problems may reflect that the agents' ability to recognise failures is larger than their ability to remedy the failures in question (ibid., pp.120-22).

The envelope model elaborates on these issue by decomposing the stage of accumulation into four substages:

(1) A *linear* stage which proceeds according to plans and budgets, because the innovation project is still premature, unified and easy to plan, and because the internal agents are eager to stick to plans and budgets in order to secure the long term survival of the project and avoid the loss of prestige and command of resources;

(2) a *ramification* stage where the project becomes increasingly complex and perhaps is diverted into subprojects that may gain a momentum of their own. The internal agents

may to an increasing degree be confronted with the requirement of controlling a chaos-like process, and the nature of their involvement may fluctuate between euphoria, frustration and despair;

(3) a *convergence* stage characterised by an increasing degree of external interference in order to unify the diversions of the project, promoting the promising ones and cut off the less promising ones by formally adjusting plans and targets; and

(4) an *implementation* stage which attempts to integrate the project in normal production activities by transferring the responsibility of the project from the initial internal agents to agents in other parts of the intraorganisational task environment. The process of transferring responsibility may be extremely difficult and even lead to a complete failure, since the logic of the innovation project has to be agreed upon by agents who employ quite different perspectives, e.g. R&D personnel, production staff and marketing people who, as described in section 1.1 and chapter 2, may find it difficult to understand and communicate across the boundaries of the rationality of each subunit.

Regarding the flexibility-stability approach advocated by the present volume, the accumulation and envelope models may be characterised as follows: In comparison to the linear models reviewed by Saren (1984) and the chain-linked model presented by Kline & Rosenberg (1986), the accumulation and envelope models appear as a synthesis in the sense that they identify from a process perspective the major (1) departmental, activity and decision stages of the innovation process, (2) stimuli-response mechanisms, (3) feedback processes and (4) interactions between different types of intraorganisational rationalities. Like the classic flexibility-stability approach developed in chapter 2 and further elaborated in chapters 3-5, the accumulation and envelope models apply a sociological point of view which integrates the technical and organisational aspects of the innovation process. They define an intraorganisational selection environment by dividing the interacting agents into internal and external agents that represent various segments of the intraorganisational task environment. However, the accumulation and envelope models differ from the approach developed in chapters 2-5 in two important respects. *First*, the main weakness of the accumulation and envelope model relates to the analysis of the post-innovation stage. Christensen, J.F. (1992, p.131) argues that neither the accumulation nor the envelope models deal explicit with the stage of take-off where the innovation project reaches the breakeven point or "either is excluded from the product mix or subjected to a renewed, ressource demanding effort that takes the form of post-innovative activities or routine-based adjustment". It may be argued that this line of criticism applies to the approach of chapters 2-5 as well. On the other hand, whether this statement is accepted or not depends on the way in which the concept of performance gaps is interpreted. As may be recalled, performance gaps are defined as a percieved deviance of performance from aspiration levels which causes the organisation members to become aware of the necessity to

> unfreeze elements within the organization most closely related to the external environmental change. When this occurs, conditions are present for altering the structure and function of the organization or some subsystem of it.
>
> (Zaltman et al., 1973, p.3)

From the evolutionary perspective described in chapter 6, the elements which may become unfreezed relate to operating procedures, investment rules or search procedures. To the extent that performance gaps occur in relation to the existing product mix, the unfreezing of elements takes the form of post-innovative activities or routine-based adjustment, e.g. through learning by using and interacting. In consequence, the learning processes initiated contribute to the accumulation of the stock of knowledge, as indicated by figures 32 and 37, and the increase of the stock of knowledge stimulates the ability of the organisation to deal with future performance gaps. The type of learning processes described by Zaltman et al. (1973) are single- and double-loop learning in the sense of Argyris & Schön (1978), and the unfreezing of elements requires that some de-learning in the sense of Johnson (1992) takes place.

Second, the accumulation and envelope models differ from the flexibility-stability approach in the way in which the element of uncertainty enters the stage. While the former models capture the distinction between incremental and radical innovation, this distinction has only been present to a very moderate extent in the flexibility-stability approach outlined so far in the present volume. Instead, uncertainty has been defined in terms of the degree to which preferred outcomes can be predicted (chapter 3); the gap between the knowledge of the organisation members and the knowledge necessary to cope with contingencies (chapter 4); the state of ambiguity, and the transition of learning in search to changes in operating procedures (chapter 6). This difference reflects that while the emphasis of the accumulation and envelope models are mainly on product innovation, the emphasis of the flexibility-stability approach has so far been mainly on organisational and process innovation. The following section 7.2 elaborates on the flexibility-stability approach regarding product innovation, uncertainty and sectoral patterns of innovation.

7.2. Patterns of innovation

Kay (1988, pp.282-83) classifies uncertainty into *general business uncertainty* about the future trend of society, *technical uncertainty* about whether or not the innovation in question will meet specific performance and cost criteria, and *market uncertainty* related to the commercial viability of the innovation. The existence of these types of uncertainties is familiar to the organisation and may, for most types of innovation, be conceptualised and handled by the existing routines of decision making, activity programmes, learning and search, at least to a degree which is satisfactory measured by the relationship between the levels of aspiration and performance. However, the types of uncertainty confronting the organisation may become unfamiliar and the outcome even more unforseeable in cases of a high degree of novelty, especially at moments where the innovation design is in a state of flux and the innovative activities of the organisation open up new technological possibilities that render the existing routines and sources of knowledge obsolete.[107] This point of view, which stresses the importance of the distinction between incremental and radical innovation, should not be interpreted as if the commercial and technical success are negatively correlated with the degree of novelty of innovation. As Adler (1989) points out, the distinction between incremental and radical innovation is too crude in many cases[108] and, furthermore,

> assumes that the greater the degree of technical newness, the lower the probability of technical success. But there may be important differences in both the degree and the type of uncertainty created by different types of newness. (...) And there is often a big difference between attacking "new to the world" and "new to the company" domains.
>
> (Adler, 1989, p.52)

Freeman (1982) provides, primarily from a producer point of view, a continuum of the degree of technical uncertainty associated with various types of innovation, ranging from "true" to "very little" uncertainty, where the degrees of uncertainty may be interpreted as varying degrees of Knightian unmeasurability (ibid., p.149).[109] As appears from figure 39, the very large degree of technical uncertainty is associated with fundamental and radical

---

107. Actually, this is what may happen when we dismiss the individual entrepreneur of the early Schumpeter and introduces the team entrepreneur of the later Schumpeter in the form of a R&D unit. As argued by Kay (1979, p.38), the role of R&D is not only that of mitigating uncertainty related to innovation, but also that of "uncertainty *generation* - finding new problems and possible solutions for future exploitation, uncovering and developing new areas of uncertainty".

108. As an example, Adler (1989) mentions the video recording industry, cf. Rosenbloom & Cusumano (1987). Lundvall (1992a) argues that incremental and radical innovations may differ in terms of the technical and economic dimensions, e.g. a technically radical innovation may be incremental in the economic dimension. Lundvall (1992a, p.12) exemplifies by mentioning "the Babbage version of the computer, obviously a radical innovation in technical terms, more than a century before it had any economic impact at all".

109. The Knightian unmeasurability amounts to the definition of uncertainty in chapter 3, see figures 15 and 16.

technical change, while major and second-generation technical change are associated with a moderate to high degree of uncertainty. The more incremental types of innovation exhibit a relatively low degree of uncertainty, but even at this part of the continuum an element of technical uncertainty is, of course, present.

Figure 39. *A continuum of technical uncertainty*

Uncertainty	Type of innovation
True	Fundamental research Fundamental invention
Very high	Radical product innovation Radical process innovation from outside firms
High	Major product innovation Radical process innovation in own establishment or system
Moderate	New generations of established products
Little	Licensed innovation Imitation of products and processes Modification of products and processes Early adoption of established process
Very little	New model Product differentiation Agency for established product innovation Late adoption of established process innovation in own establishment Minor technical improvements

Source: Freeman (1982, p.150), table 7.1

Like Nelson & Winter (1977), Freeman (1982) searches for a useful theory of innovation and argues that

> other ways of interpreting and understanding innovative behaviour are needed (Nelson and Winter, 1977 and 1982). One possible approach to such a theory (and it is no more than a first approach) is to look at the various *strategies* open to a firm when confronted with technical change.
>
> (Freeman, 1982, p.169)[110]

---

110. Freeman's reference to Nelson & Winter (1982) is Nelson & Winter (1982a).

This *first approach* is presented as a typology of six types of strategies which "should be considered as a spectrum of possibilities, not as clearly definable pure forms" (ibid., p.170). These strategies are presented in figure 40 which reflects that each strategy emphasises different types of economic activities and that the overall innovation strategy is a mix of in-house activities. The nature of this mix depends on the degree of uncertainty related to the different types of activities.[111] A *second approach*, which in effect elaborates on the Freeman approach, might be to supplement the analysis of the match between the various activities and the strategic behaviour of the organisation with an analysis of the match between the environment and the intraorganisational context. This symbiotic approach is reflected in Burgelman & Rosenbloom (1989) who focus upon the interplay between the forces *generating* and the forces *integrating* the technical capabilities (competencies) of the organisation. They argue that

> successful firms operate within some sort of harmonious equilibrium of these forces. Major change in one, as in the emergence of a technological discontinuity, must be matched by adaptation in the others.
>
> (Burgelman & Rosenbloom, 1989, p.20)

× matching

Burgelman & Rosenbloom (1989), who are inspired by a *cultural evolutionary* perspective according to which strategy making is "a social learning process", propose a capabilities-based perspective on technology strategy. This perspective describes (1) the strategic behaviour of organisations and the technological evolution of the economic environment as mechanisms that generate organisational capabilities, and (2) the structure of the organisation and its environment as mechanisms that integrate capabilities. The *generation* of technical capabilities is moulded by organisational experience, reflected in routines for strategy making and existing capabilities, and it depends on the evolution of dominant technical designs and regimes in the economic environment. The *integration* of the capabilities generated depends on the intraorganisational context that serves as "an internal selection mechanism" (ibid., p.7), and on the external selection mechanism embedded in the environmental context. From this perspective, the organisation becomes a nexus between interacting internal and external forces in the sense that organisational

cultural evolutionary

---

111. However, there is, of course, not a one-to-one correspondence between the degree of uncertainty and the relevant technology strategy, since the choice of technology may depend on a number of factors such as the appropriability regime, the complementary assets and the dominant design paradigm associated with the innovation in question (Teece, 1986), all of which contribute to the general business and market uncertainty supplementing the technical uncertainty faced by the organisation.

Figure 40. *The relative importance of various in-house scientific and technical functions*

Activity	Strategy					
	Offen-sive	Defen-sive	Imita-tive	Depen-dent	Tradi-tional	Oppor-tunist
Fundamental research	4	2	1	1	1	1
Applied research	5	3	2	1	1	1
Experimental development	5	5	3	2	1	1
Design engineering	5	5	4	3	1	1
Production engineering, quality control	4	4	5	5	5	1
Technical services	5	3	2	1	1	1
Patents	5	4	2	1	1	1
Scientific and technical information	4	5	5	3	1	5
Education and training	5	4	3	3	1	1
Long-range forecasting, product planning	5	4	3	2	1	5

Range 1-5 indicating weak (or non-existent) to very strong

Source: Freeman (1992, p.171), table 8.1

behaviour is moulded by the external selection environment which is, on the other hand, moulded by the organisational activities effectively carried out.

However, the integration of capabilities is by no means a simple task. As proposed by the simple model and discussed in chapters 2-6, the internal selection environment may be characterised by conflicting rationalities in the way proposed by the accumulation and envelope models. For instance, in developing a new product the R&D unit may be concerned with incorporating the latest technical advance and pay less attention to the technical capabilities of the production unit. The production unit, on the other hand, may focus on product improvements that utilise the existing capabilities at the technical core. This is an example of the differences in time horisons proposed by Woodward (1965), cf. chapter 1. If communication between these two groups are infrequent and punctuated, the activities of the R&D unit may gain a momentum of its own, and the ensuing innovation proposal from the R&D unit may pay less attention to manufacturability than to technical

,tication. In consequence, the innovation proposal may be opposed by the production unit, and conflict is bound to become tense if, for instance, marketing judges the market opportunities different from the R&D and production units. In consequence, the innovation proposal creates difficulties in the process of manufacturing, and although ensuing adjustments may be undertaken by the production unit, marketing of the product may fail because the link between the market needs and the product becomes blurred.

This case may in some instances take the form of a *relay race* where the innovation is pushed from one department to another during the innovation. The relay race is linear in the sense that the feedback mechanisms stressed by the chain-linked model are absent. The opposite case, in which a product champion or a project group is assigned to bridge the conflicting rationalities, may be described by the metaphor of a *rugby team* where each player acts according to a common and formally pronounced strategy. Reviewing a number of industries, Pay (1990) uses the terms of *relay race* and *rugby team* in order to describe the innovation process of, respectively, American and Japanese firms, while German firms is found to be located somewhere in between. In consequence, the time consumed during the innovation process and the costs involved seem to be higher in American than in Japanese firms, and less in Germany than in the US. The empirical evidence presented by Pay (1990) seems to indicate that this proposition holds for the electrical industry regarding lead time and innovation costs, and the automobile and chemical industries regarding lead time.[112] The evidence can to some extent be explained by the relative emphasis placed on interdepartmental communication and cooperation.

Like in the work of Freeman (1982), the implication of Pay (1990) is that the innovation process may be described in terms of the *focus* of the overall innovation strategy. This theme is pursued by Cooper (1983) who employs an activity-stage approach in order to characterise the product innovation process of 30 industrial firms comprising 58 innovation projects. As table 3 shows, no typical innovation process model is identified but instead seven activity-stage models emerge. The activity within each model is described in terms of marketing, evaluation of the project, and the technical/production activity associated with design, engineering and production, and the approach adopted by Cooper (1983) focusses on the extended process of innovation depicted in figure 37.

The *market oriented* innovation process is characterised by a lengthy period of customer trial of a protoptype and a lengthy market launch phase of the final product variant. Innovation proposals are primarily derived from market needs, and the firms involved is characterised by low R&D spending, but strong management and marketing

---

112. In fact, Germany proved to have a competitive advantage in the machinery and metals industry regarding both lead time and innovation costs.

capabilities. Market research activities are undertaken simultaneously with product design and development, and prototype trials by costumers coincide with design and development activities and with the final market launch.

Table 3. *Shares of total time of product innovation projects spent on major activity-stages, assessed by respondents, percentages, N = number of projects*

Process	Activity			Total	N	Success rate (1)
	Marketing	Technical/ production	Evaluation			
Market oriented (2)	73.7	24.8	1.4	99.9	11	52
Design dominated	28.3	71.7	0.0	100.0	10	40
Balanced complete	31.3	55.0	13.7	100.0	7	71
Front-end dominated	57.6	42.0	0.3	99.9	8	50
Minimum	71.0	28.2	0.8	100.0	9	44
Launch with prototype	18.8	78.8	2.4	100.0	6	50
Prototype dominated	23.2	76.7	0.1	100.0	7	71
Average (3)	46.3	51.3	2.3	99.9	58	52

(1) Share of projects assessed as a success.
(2) Success rate assessed from the description of the cluster in the source.
(3) Activities and total are calculated as weighted average from the observations in the table, since detailed information is not given by the source.

Source: Cooper (1983)

The *design dominated* process is, of course, characterised by a lengthy design phase, and while the innovation proposal in the market oriented process is derived from market needs, the design dominated process combines assessments of market needs and the generation of ideas from technological opportunities. A relatively small amount of time is spent on marketing, and the process is dominated by trial production and the acquisition of production facilities. During this process, marketing activities are undertaken and coincide with trial production and the acquisition of production facilities.

The *balanced complete* process is characterised by a balance between marketing and the technical/production activities throughout the duration of the process. However, the stage of screening (searching) for ideas is significantly longer than in the other models and dominates, especially the early phases of the innovation process. The firms in question spend relatively more on R&D than other firms and are, seemingly, more successful.

The *front-end dominated* process emphasises the combination of market derived and technological idea generation based on preliminary market forecasting and a strong commitment to the pre-design (front-end) stages.[113]

The *minimum* process is, primarily, a two-stage process dominated by a product design phase followed by a market launch phase that is significantly longer than in the other types of processes. Market idea generation is, on the other hand, significantly shorter and firms that engage in the minimum process are, on average, less successful, perhaps because they do not differentiate themselves from other firms in terms of heavy R&D spending, strong technological capabilities and the like.

The *launch with prototype* process distinguishes itself by an early market launch that follows upon a design phase significantly longer than in the other types of processes. Product design and development dominates the processes, prototype contruction endures much longer, and market launch appears to coincide with in-house prototype testing. In terms of management, marketing and technical capabilities, the firms which engage themselves in this type of project are relatively weak.

Finally, the *prototype dominated* process is characterised by in-house prototype testing that consumes more than 2/3 of the innovation time. Activities related to design and testing are undertaken in parallel. The market launch stage is rather short since firms that employ this kind of process tend to have strong management and marketing capabilities and, thus, is relatively successful.

Two conclusions emerge from the empirical evidence supplied by Cooper (1983). *First*, the innovation process is characterised by an overlap between its various phases and is far from sequential in nature, as discussed in the previous section. *Second*, the most successful firms are those which are able to undertake a variety of activities and to

---

113. "Front-end dominated projects were unique in that they were not specific to any particular type of firm" (Cooper, 1983, p.10).

balance market oriented and technical/production oriented activities; i.e. the most successful firms are those who, in practice, avoid the technology-push/demand-pull dichotomy. In conclusion, it seems reasonable to argue that a balance between market oriented and technical/production oriented activities are crucial to the solution of the innovation design dilemma.

The seminal work by Rothwell et al. (1974) provides some insight into how the balance between market oriented and technical/production oriented activities may be achieved through a solution to the flexibility-stability dilemma. Rothwell et al. (1974) focus on fourty-three pairs of success and failure of innovations in chemicals and scientific instruments and point to especially five factors that distinguish successful firms and innovations from unsuccessful ones: Better understanding of user needs, more attention to marketing and publicity, more efficient development work, more use of outside technology and scientific advice, and more authority and experience invested in those in charge of the innovation process. They suggest that the decision to innovate in successful firms are made more for commercial reasons than out of consideration for technical success, and the innovation proposal seems to encounter relatively small resistance on commercial and technical grounds. Innovation is an important part of the company strategy, and large teams are employed at the beginning of development. The R&D effort is based more on a sophisticated understanding of user needs than on high expectations regarding technical success.[114] Innovative efforts rely to an important extent on external sources of knowledge and technology. Internal communication seems to be highly efficient, and those in charge of the innovation process enjoy a fairly amount of power, responsibility, experience, enthusiasm and prestige.

In sum, the preceeding discussion points to the successful innovation project as one where (1) the phases of the innovation process tend to overlap, (2) the organisation provides effective screening for new ideas, and (3) resources are devoted to intraorganisational communication, manning, attention and authority. However, this is by no means a simple task in the conduct of daily organisational life. *First*, as chapters 2-6 have shown, the nature of behavioural regularities in relation to organisational action may impede the achievement of a successful outcome of the innovation project. Chapter 8 returns to this issue. *Second*, the nature of the innovation process may not only differ across firms, as

---

114. Successful firms percieve user needs at an earlier stage than unsuccessful firms, seemingly because they have a better understanding of user needs. Similarly, successful innovations are better adapted to user needs and require fewer modifications as users learn to employ the innovation. Thus, learning by using and learning by interacting seems to be more effective in the case of successful innovations than in the case of unsuccessful ones. Furthermore, successful firms tend to devote more effort to overt publicity and to the education of users.

discussed above. It is also possible to distinguish between sectoral differences that define boundaries around the innovation process within each firm that belongs to the sectors in question.

Employing the innovating firm as the basic unit of analysis, Pavitt (1984) has suggested a sectoral classification with four groups: *Supplier-dominated* firms, typically found within sectors like agriculture and traditional manufacturing; *scale-intensive* firms, typically within the sectors of bulk materials and assembly; *specialised equipment suppliers*, e.g. machinery and instruments; and *science-based* firms within sectors like electronics and chemicals.[115] Each sector is distinguished according to (1) technological opportunities in terms of the sources of technology, (2) demand conditions in terms of how the users valuate production costs *vis-à-vis* technical performance, and (3) the means of appropriability in terms of how the innovating firms seek to appropriate the benefits from innovation. Figure 41 summarises the relationship between firms within the four categories in terms of the three distinguishing factors. The circles describe the demand conditions in terms of the user's primary focus of attention, and the price/cost circle indicates that the products of the users in question are price-sensitive; thus, the users are focused on productivity gains as a means to reduce production costs. This focus of attention characterises especially the supplier-dominated and scale-intensive firms. The performance circle indicates that the users are more oriented towards technical reliability and performance than towards the price/cost relationship, and this focus of attention characterises mainly specialised equipment suppliers. Finally, science-based firms represents a mix due to the existence of a

Figure 41. *Sectoral patterns of innovation*

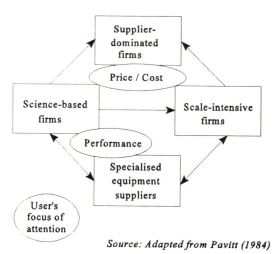

Source: Adapted from Pavitt (1984)

---

115. The scale-intensive firms and the specialised suppliers may also be characterised as production-intensive. The Pavitt classification is based on about 2,000 innovations included in the innovation data bank at the Science Policy Research Unit at University of Sussex. The data bank is described in Townsend et al. (1981).

wide variance in relative emphasis on production and process technology within each of the sectors, reflecting the different cost/performance trade-off for consumer goods, standard materials and specialised professional applications.

(Pavitt, 1984, p.362)

The four groups of innovating firms differ according to the technological opportunities they exploit and the means of appropriability they use. The sources of technology in the innovative *supplier-dominated* firms are mainly the suppliers of the firm (as the label of the group indicates). The relative size of the firms is small[116], and the innovative process is biased towards process innovation focused on cost-cutting.[117] The means of appropriability are mainly non-technical, e.g. marketing and design, and learning by doing is an important source of the improvement of productivity.

The innovative *scale-intensive* firms are relatively large and match specialised equipment suppliers with strong in-house capabilities as sources of technology. The innovative effort is biased towards process innovation focused on cost-cutting, and the means of appropriability relate to the technical capabilities of the firm, primarily protected through secrecy, patents and technical lags. Learning by doing is, obviously, important, but the scale-intensive firms match learning by doing with the exploitation of dynamic learning economies associated with learning by interacting and the development of strong in-house capabilities.

As one would expect, the innovative effort by *specialised equipment suppliers* are biased towards product innovation with an emphasis on product design. The firms are relatively small, and important means of appropriability are, besides patents and design knowhow, the knowledge of users exploited through learning by using. The specialised suppliers match the knowledge of customers with strong in-house capabilities as sources of technology.

---

116. The relative size is measured in terms of the number of employees, cf. the following table.

The size distribution of the Pavitt (1984) sample, expressed in percentages, shows a bias towards very large firms, indicating that the innovations recorded in the SPRU data bank reflects mainly the efforts of firms with a large economic impact on the total amount of innovations in the British economy.

No. of employees	Total sample
1-999	24,9 %
1,000-9,999	21,9 %
10,000 +	53,1 %

117. Product innovations are defined as "those innovations that are used *outside* their sector of production" and process innovations as "those that are used *inside* their sector" (Pavitt, 1984, p.348).

Finally, the innovative *science-based* firms resemble the scale-intensive firms in the sense that they are relatively large and match the knowledge of suppliers with strong in-house capabilities as the primary sources of technology. In consequence, learning by interacting is important to the development of technical capabilities, and the production engineering and R&D departments are important sources of technological knowledge, as in the case of the scale-intensive firms. However, the science-based firms differentiate themselves in that they *are* science-based, and R&D knowhow is, in consequence, more important as a means of appropriation than in any other types of firms. The innovative effort is mixed, although with a small bias towards product innovation.

The approaches underlying the work of Cooper (1983) and Pavitt (1984) are similar in that they take, as their point of departure, the innovating firm from a cross-sectional sample. On the other hand, there are important differences in the way in which these approaches are utilised for analytical purposes, since Cooper (1983) identifies seven clusters, i.e. types of innovation processes, which cannot be grouped according to sectors, while Pavitt (1984) identifies four clusters grouped according to sectoral affiliation.[118] However, is is not impossible to reconcile the results obtained. The Cooper (1983) classification identifies the main activity stages of the innovation process in a way which determines how the various activity stages overlap according to the capabilities that characterise the firms within each cluster. The Pavitt (1984) classification explains how these capabilities are linked with technological opportunities, demand conditions and means of appropriability. In consequence, it would be possible to determine the distribution of the Cooper-clusters across the Pavitt-categories through further studies. Following Freeman (1982), one might argue that although the innovative activities of a firm is located within the boundaries prescribed by technological and market possibilities, there are some degrees of freedom which reflect the opportunities of the firm to influence the technological and market possibilities, i.e. the innovation process is characterised by an interplay between the intraorganisational and extraorganisational task environments, to put it in evolutionary terms. As Freeman (1982) argues:

> Within these limits, the firm has a range of options and alternative strategies. It can use resources and scientific and technical skills in a variety of different combinations. It can give greater or lesser weight to short-term or long-term considerations. It can form alliances of various kinds. It can license innovations made elsewhere. It can attempt market

---

118. Of course, an important reason for why the Cooper (1983) sample cannot be grouped according to sectors is that the sample comprises only 30 industrial product firms. However, this is actually a fairly large sample, since the data employed in Cooper (1983) were obtained through in-depth interviews.

and technological forecasting. It can attempt to develop a variety of new processes and products of its own. It can modify world science and technology to a small extent, but it cannot predict accurately the outcome of its innovative efforts or those of its competitors, so that the hazards and risks which it faces if it attempts any major change in world technology are very great.

(Freeman, 1982, p.169)

This point of view was previously described as a *first approach*, as reflected in figure 40. While the symbiotic approach of Burgelman & Rosenbloom (1989) was described as a *second approach*, the analysis of how the Cooper (1983)-clusters are distributed across the Pavitt (1984)-categories may, accordingly, be described as a *third approach*. However, an elaboration of this kind lies outside the scope of the present study. Instead, the preceding section moves from the level of the innovating firm and its sectoral affiliation to the macro level of the innovation process and discusses the possibility of identifying some stylised facts of innovation, in general and applicable to the analysis of technical and organisational innovation during the present decade.

7.3. Technological paradigms and techno-economic changes

As has appeared from the discussions in the present volume, the flexibility-stability approach is applicable to organisational innovation, technical innovation (product and process) and to the interplay between organisational and technical innovation. In pursuit of these use values of the theory, the flexibility-stability approach has been elaborated both in terms of studies based on empirical findings *and* generalising theory. It has been obvious to adopt this combined approach, since it characterises the line of thinking within the organisation theories that are the origin of the flexibility-stability approach. However, as far as the field of technical change is concerned, a combined approach is less obvious, since the field of technical change has developed within economics where the phenomenon of technical change most often enters the analysis as an *exogenous* or *endogenous* cause of economic growth. Although technical change sometimes are treated as partly endogenous and partly exogenous in the same model, economists in general agree on the fact that we may speak of two competing approaches. This dichotomy may be explained by differences in the perception of what economic analysis is about.

Nelson & Winter (1982) distinguish between appreciative and formal theory. While appreciative theory attends to empirical findings and embarks on story-telling in terms of a causal interpretation of the evidence at hand, formal theory operates at some distance from the empirical evidence and aims to find and explore logical relationships which might push the borders of analytical thinking. Appreciative and formal theory do not

175

develop independently of each other, but the relationship between them is rather complicated:

> Empirical findings or facts seldom influence theorizing directly. Rather, in the first instance they influence appreciative theorizing. In turn appreciative theorizing provides challenges to formal theory to encompass the understandings of appreciative theory in stylized form. The attempt to do so may identify gaps or inconsistencies in the verbal stories, and often may suggest new appreciative theory storylines to explore. In turn the empirical research enterprise is reoriented.
>
> According to this view, formal theory can lead appreciative theory, taking the research enterprise into new arenas or in new directions. Or it can follow, correcting, fine tuning, appreciative theory. But it can also proceed mainly on its own with little contact with the rest of the research enterprise, neither extending it or correcting it.
>
> (Nelson, 1992, p.5)

Nelson (1992) describes the development of economic growth theory during the afterwar period in terms of the relationship between appreciative and formal theory, and suggests that an important part of the difficulties encountered by growth theory in order to explain the slowdown of economic growth during the 1970s-80s recession might be accounted for by an increasing divergence between appreciative and formal theory. Whether future research will remedy this divergence or not is, of course, an open question, and it seems as if Nelson (1992) is sceptical towards the ability of contemporary formal theory to encompass the lessons from contemporary appreciative theory since he argues that

> the main thing behind what has been the matter with formal growth theorizing since the postwar rebirth of research in economic growth has been that formal theorists have not tried very hard to understand and build into their theorizing what has already been in appreciative theory. The new generation of growth theorists are doing somewhat better in this regard. The main thing I see as potentially the matter with the new wave of formal growth theorizing is that the theorists may have a proclivity to stay in the old intellectual grooves, or ruts, not recognizing the cutting edge of analysis may now be on topics that have not traditionally been adressed in modern economics.
>
> (Nelson, 1992, pp.45-46)

The cutting edge of contemporary appreciative analysis is on technological advance in terms of the relationship between the firm and its environment, often with references to various types of institutional arrangements such as the R&D system, the system of regulation and legislation, the social values of the economic agents, the system of production, and the innovation system. At the core of this type of appreciative theorising is the same belief which underlies the discussions of the present volume, i.e. that "technologies and business firms should be understood as co-evolving" (ibid., pp.41-42).

One line of criticism of the formal growth theorising, which appears to emanate from this type of appreciative theorising, is that the theorist has to abandon the conventional assumption that technology can be described in terms of a set of blueprints, and that technical progress may be measured as additions to the available stock of blueprints that any firm is able to implement in production. Instead, one must recognise that technological change depends on the *technical capabilities* of the firm:

> Mastery of technique is organizational rather than individual in several respects. The practice of complex technologies inherently involves organization and management. The way firms organize to implement a common broad technology can make an enormous difference. This is a key finding of most of the recent work comparing U.S. and Japanese auto production. Further the competence resides in the organization and not simply in the set of individuals.
>
> (Nelson, 1992, p.36)

At the level of the stylised facts of innovation, which is the topic of this last section of chapter 7, there has been several attempts to provide a general classification within the lines suggested by Nelson (1992). These attempts share the basic analytical property that they distinguish themselves from the more orthodox economic reasoning on technical advance. In traditional economic analysis of technical advance, which focusses on the economisation of scarce resources in order to obtain increasingly higher levels of output, the definition of innovation is normally based on the distinction between product and process innovation where the latter is at the centre of the analysis and presented as "an addition to the existing technical knowledge" (Blaug, 1974, p.495) or "the advancement of knowledge about methods of production" (Hacche, 1979, p.100).[119] In the attempts within the field of innovation economics to provide some stylised facts of innovation, the tendency to focus on one type of innovation is less clear, although there seems to be a preponderance to focus on product innovation. This may reflect, partly, the point of view that process innovation appears as the diffusion of product innovation, as in the case of the Pavitt (1984) analysis, and partly a division of analytical labour regarding the process of diffusion and the process of innovation generation.[120]

How should the stylised facts of innovation be assembled and described? Dosi (1988) suggests that innovation, essentially,

---

119. Product innovation is, however, at the centre of the analysis in cases where the traditional economic analysis is preoccupied with diversions from perfect competition.

120. For instance, while the work by Rogers (1983) clearly reflects the diffusion-oriented approach, the chain-linked model reflects an innovation-generation approach.

concerns the search for, and the discovery, experimentation, development, imitation, and adoption of new products, new production processes and new organisational set-ups.

(Dosi, 1988, p.222)

Thus, the stylised facts of innovation should reflect all of these processes, and, consequently, emphasise the systemic nature of the innovative process. Within such a systemic framework, Dosi (1988, pp.222-23) presents five stylised facts of innovation, i.e.

(1)   *uncertainty*, in the sense that outcomes to preferred actions are difficult to predict, and that there may exist a number of alternatives not yet discovered;

(2)   *reliance on advances in scientific knowledge*, which points to the increasing importance of novel opportunities created by science;

(3)   *institutionalisation of search activities*, i.e. the incorporation of search in formal organisations;

(4)   *learning*, in the sense described in section 6.3; and

(5)   *cumulativeness*:

...it seems that the patterns of technological change cannot be described as simple and flexible reactions to changes in market conditions: (i) in spite of significant variations with regard to specific innovations, it seems that the directions of technical change are often defined by the state-of-the-art of the technologies already in use; (ii) quite often, it is the nature of technologies themselves that determines the range within which products and processes can adjust to changing economic conditions; and (iii) it is generally the case that the probability of making technological advances in firms, organisations and often countries, is among other things, a function of the technological levels already achieved by them.

(Dosi, 1988, p.223)

The notion of cumulativeness suggests that innovation tends to follow certain paths which combine the occurrence of new opportunities with the existence of restrictions on the process. This situation may, theoretically, be interpreted in a number of ways.

First, one may take the point of view that cumulativeness appears as the consequence of previous investments and innovative efforts that may be regarded as sunk costs. In order to ripe the fruits of previous endeavours, the economic agents are inclined to continue their productive efforts along these lines.

Second, one may argue that a specific path of innovation tends to become self-reinforcing as the economic agents discover new opportunities on the basis of previous accumulated knowledge.

Third, the interaction between economic agents may favour some types of economic activities and not others, since the economic benefits from that interaction depends on the complementarity between the activities of each agent.

This line of reasoning seems to underly the argument of Arthur (1988) that the attractiveness of some innovation depends on "increasing returns to adoption". According to Arthur (1988), who focusses on technical innovation, increasing returns to adoption stems from a combination of learning processes and economic benefits. The development and improvement of a technology is positively correlated to the rate of adoption due to the existence of learning by using. Furthermore, the availability of new opportunities associated with that adoption increases as the technology spreads throughout the economic environment, partly because the producers of that technology tries to serve an increasing market, and partly because these opportunities enters as part of the infrastructure of the technology in question. Finally, increasing returns to scale may be at work, both in the sense of scale economies in production and informational increasing returns.[121] In consequence, the relative advantage of the technology increases.[122]

Within a more systemic framework, the increase of the relative advantage of some technology has been stated by Freeman & Perez (1988) in terms of the concept of *techno-economic paradigms*. Based on the empirical work at the Science Policy Research Unit at the University of Sussex, Freeman & Perez (1988) suggest a taxonomy of innovations, which in the words of Christensen, J.F. (1992, p.54) is hierarchial and cumulative. The taxonomy refers to the degree by which the innovation in question presents something new to the state-of-the-art, and describes innovation in terms of incrementality and radicality, changes of technology systems, and changes in technoeconomic paradigms.

Changes in technology systems, which rest on a combination of incremental and radical innovation, take place at the interfaces between sectors, affect not only the sectors involved, but actually contribute to the creation of entirely new sectors. The co-evolution of existing and new sectors is based on changes in the way productive activities are

---

121. Regarding informational increasing returns, Arthur (1988, p.591) argues that often "a technology that is more adopted enjoys the advantage of being better known and better understood. For the risk-averse, adopting it becomes more attractive if it is more widespread".

122. The use in the present section of the notion of relative advance has been borrowed from Rogers (1983), to whom relative advance signifies "the degree to which an innovation is percieved as better than the idea it supersedes" (ibid., p.15). In his analysis of diffusion, Rogers (1983) suggests that the diffusion of innovation is, furthermore, influenced by the compatibility of the innovation with the values, experience and needs of the adopter, and the adopter's opportunity to understand the innovation, e.g. through experimentation or observation of the results from the use of other adopters. Rogers (1983) uses the concepts of compatibility, complexity, trialability and observability to coin these processes.

179

coordinated and organised. Freeman & Perez (1988, p.46) refer to organisational and managerial innovations that affect "more than one or a few firms", and following the approach of Freeman (1992a), these types of organisational and managerial innovation make take place within the affected firms and sectors without altering the rules of the game of the economic system as such:

> Some matching of institutions and technology may take place locally through organisational innovations at the level of the firm or group of firms. However, the bigger changes in technology may lead to a wider institutional change.
>
> (Freeman, 1992a, p.132)

Freeman & Perez (1988) associate these "bigger changes in technology" with the Schumpeterian notion of technological revolutions based on clusters of incremental and radical innovations that create a number of new technology systems.

> A vital characteristic of this fourth type of technical change is that it has *pervasive* effects throughout the economy, i.e. it not only leads to the emergence of a new range of products, services, systems and industries in its own right; it also affects directly or indirectly almost every other branch of the economy, i.e. it is a 'meta-paradigm'.
>
> (Freeman & Perez, 1988, p.47)

According to Perez (1983), this meta-paradigm may be described as a techno-economic paradigm in the sense that the technological revolution implies a new match between the economic and social systems. Propelling the advent of a new techno-economic paradigm is some key factor that may be used "in many products and processes throughout the economic system either directly or (more commonly) through a set of related technical and organisational innovations, which both reduce the cost and change the quality of capital equipment, labour inputs, and other inputs to the systems" (Freeman, 1992a, p.135). The diffusion of this key factor is facilitated by decreasing relative prices and increasing supply of the factor in question. The diffusion of the key factor is associated with the growth of main carrier branches and sectors which experience high growth rates induced by the diffusion of the key factor.

Freeman & Perez (1988) distinguish between successive *Kondratieff* periods and suggest that the Schumpeterian long cycles of growth and decline may be characterised in terms of such key factors, cf. figure 42. For instance, while the third Kondratieff was characterised by steel as the key factor carried by the growth of such new sectors as electrical engineering and electrical machinery, the fourth Kondratieff was characterised by energy as the key factor carried by the growth of sectors devoted to transportation, consumer durables and various types of process production. The new key factor, energy, assisted in overcoming some of the limitations of the third Kondratieff associated with batch production through "flow processes and assembly-line production techniques, full standardisation of components and materials" (ibid., p.52). During the fourth Kondratieff, the competitive conditions changed through the increasing occurrence of oligopolistic competition, internationalisation and arms-length subcontracting at the interorganisational level, and the advent of divisionalisation and hierarchial control.

Figure 42. *Technical and organisational properties of successive Kondratieff periods*

Periods	Description	Key factor	Limitations of previous paradigm
1.Kondratieff 1770/80s-1830/40s	Early mechanisation	Cotton Pig iron	Scale, process control, mechanisation and hand-operated tools
2. Kondratieff 1830/40s-1880/90s	Steam power and railways	Coal Transport	Location, reliability and scale of production *vis-à-vis* water power
3.Kondratieff 1880/90s-1930/40s	Electrical and heavy engineering	Steel	Use value of iron and water powered production techniques
4.Kondratrieff 1930/40s-1980/90s	Fordist mass production	Energy	Scale of batch production
5.Kondratieff 1980/90s-?	Information and communication	Micro-electronics	Diseconomies of scale, inflexible production equipment, high energy and materials intensity

Source: Freeman & Perez (1988, pp.50-53), table 3.1

Transition

This description of the transition from one Kondratieff period to another indicates that the techno-economic paradigm of each period is characterised by (1) a number of opportunities to overcome the limitations of the previous paradigm, (2) an increasing level of exploitation of the opportunities entailed in the present paradigm, and (3) a decreasing rate of benefits obtained from the productive resources, as the opportunities become fully exploited.[123] Freeman & Perez (1988) argue that the present Kondratieff period has

---

123. As Dosi (1988, p.229) points out, the occurrence of "diminishing returns to innovative efforts within the limits of a *specific* paradigm" is known as Wolff's Law. The work of Perez & Soete (1988) provide an example of how Wolff's Law may be interpreted in the analysis of international diffusion of new technology in terms of the debate on the catching-up effect.

entered the third stage, as economic activities based on high energy-intensity and dedicated production equipment become increasingly less profitable compared to the opportunities offered by technical and organisational innovation based on the diffusion and implementation of microelectronics and information technology. They characterise the mass production fourth Kondratieff in the following way:

The ideal — one best syste [handwritten]

> Its 'ideal-type' of productive organisation at the plant level was the continuous-flow assembly-line turning out massive quantities of identical units. The 'ideal' type of firm was the 'corporation' with a seperate and complex hierarchial managerial and administrative structure, including in-house R & D and operating in olipolistic markets in which advertising and marketing activities played a major role. It required large numbers of middle-range skills in both the blue- and white-collar areas, leading to a characteristic pattern of occupations and income distribution. The massive expansion of the market for consumer durables was facilitated by this pattern, as well as by social changes and adaptation of the financial system, which permitted the growth of 'hire purchase' and other types of consumer credit. The paradigm required a vast infrastructural network of motorways, service stations, airports, oil and petrol distribution systems, which was promoted by public investment on a large scale already in the 1930s, but more massively in the post-war period. At various times in different countries both civil and military expenditures of governments played a very important part in stimulating aggregate demand, and a specific pattern of demand for automobiles, weapons, consumer durables, synthetic materials and petroleum products.
>
> (Freeman & Perez, 1988, p.60)

The advent of microelectronics and information technology removes the logic of productive affairs from the energy- and material-intensive arena towards another setting, where competitive advantages are obtained through information-intensive activities. The productive affairs of this new setting are characterised by less dedicated and thus more flexible production systems which render scale economies less important, for a number of reasons, which are elaborated in chapter 8.

First, a number of new opportunities arise with respect to the economisation of inputs, as argued by Sundqvist et al. (1988). The economisation of inputs relates to: Continous improvement of products and processes through on-line observation and quality control; increased levels of tolerance and flexiblity of production equipment; decreased numbers of electromechanical parts and processes, and thus a decreased number of possible defects in assembly; and the opportunity of delegating areas of authority over production activities. In consequence, fewer products are rejected, the amount of work-in-progress reduced, and the reliability of the end product increased. This implies that the input to the production process may be characterised by considerable savings in capital, energy, material and labour.

Scale economics. less important — Fredysn of misnath / Sif ? Kennot? [handwritten]

182

Second, due to the programmability of the new production equipment and the economisation of inputs, the size of profitable production series tends to decrease and the size of profitable series variation tends to increase, as observed by Bessant (1989). This phenomenon, often referred to as *economies of scope*, implies that the firms which exploit the opportunities of microelectronics and information technology may to a larger extent be able to serve markets that are less standardised and more costumised than during the mass production Kondratieff and thus to a larger extent exposed to short product life cycles. Expressed in evolutionary terms, these firms are gradually altering their productive routines, thus changing the internal selection environment and, during this process, providing a stimulus to the external selection environment in the same direction as the change of productive routines. As the outcome of the co-evolution of firms and markets, the new information and communication Kondratieff gradually gains a momentum of its own due to cumulativeness in the sense of Arthur (1988) described previously.

This line of argument has increasingly been applied by scholars in the field of innovation economics. This is, of course, a result of the accumulation of empirical evidence, but it may also be interpreted as the result of an increasing analytical adherence to the Schumpeterian idea of technological revolutions, which has become revitalised by the concept of technoeconomic paradigms. Originally, the concept of technological paradigms was introduced by Dosi (1982) who compared this concept to the Kuhnian perception of scientific paradigms and defined a technological paradigm as comprising positive and negative heuristics, i.e.

> strong prescriptions on the *directions* of technical change to pursue and those to neglect. (...) Technological paradigms have a powerful *exclusion* effect: the efforts and the technological imagination of engineers and of the organizations they are in are focussed in rather precise directions while they are, so to speak, "blind" with respect to other technological possibilities.
>
> (Dosi, 1982, pp.152-53)

The Dosian technological paradigm may be characterised as a subset of the technoeconomic paradigm, since the latter "go beyond engineering trajectories for specific product or process technologies and affect the input cost structure and conditions of production and distribution throughout the system" (Freeman & Perez, 1988, p.47).[124] It is from this

---

124. The term "trajectory", which was originally proposed by Nelson & Winter (1977), refers to Dosi's definition of a technological trajectory as "the pattern of 'normal' problem solving activity (i.e. of 'progress') on the ground of a technological paradigm" (Dosi, 1982, p.152), or "the activity of technological progress along the economic and technological trade-offs defined by a paradigm" (Dosi, 1988, p.225).

183

analytical perspective of technological and technoeconomic paradigms that the next chapter turns to some evidence on the flexibility-stability dilemma.

Chapter 8

Implementation of High Technology

8.1. Defining advanced manufacturing technology

The Schumpeterian approach of Freeman & Perez (1988) presented in the previous section implies that the industrialised economies are gradually entering a period of technical and organisational development characterised by the diffusion of microelectronics and information technology. According to this approach the comparatively most competitive firms are those which apply a systemic approach to the integration of technical and organisational change based on advanced manufacturing technology. From an evolutionary perspective, the diffusion and implementation of advanced manufacturing technology is caused by, and causes, a change of the interplay between the internal and external selection environments of the firm. Furthermore, a systemic approach may, on the one hand, overcome some of the problems entailed in the flexibility-stability dilemma, but are, on the other hand, likely to result in some new types of problems. This dynamic perspective is adressed in the present chapter.

Advanced manufacturing technology is frequently associated with various types of technology which incorporate microelectronics and information technology. Following Zairi (1992), advanced manufacturing technology refers to the programmable automation of productive processes which involves the use of computerised devices. Programmable automation may be described with reference to two main groups of manufacturing activities, i.e. computer aided manufacturing and computer aided design and engineering. Computer aided manufacturing involves the use of numerically controlled machine tools, for instance in the formation of flexible manufacturing systems that brings economies of scale to batch work, and the use of computerised planning systems such as computer aided process planning and manufacturing resource planning. These types of systems may be combined with automatic material handling systems and could take the form of group technology. The compatibility of the constituent parts of an advanced manufacturing technology system depends on the way in which information flows is directed and coordinated, and thus it would seem natural to include in the set of techniques such information controlling protocols as the manufacturing automation protocol developed by General Motors, or the technical office protocol developed by Boeing. The control system involved in advanced manufacturing technology systems is typically described with reference to control philosophies, such as *just-in-time*, which refers to inventory control, *kanban*, which refers to the integration of just-in-time principles and procedures for avoiding operational defects, and *computer integrated manufacturing*, which refers to the integration of the various types of techniques, planning systems and protocols.

This general description contains what Zairi (1992) terms *the mechanistic aspects* of advanced manufacturing technology. However, in relation to the flexibility-stability dilemma the interesting aspects of advanced manufacturing technology is not the specific types of technology or technology systems applied, but the implications for the way in which the utilisation of these technologies are organised. This implies that the present study should adopt a more comprehensive definition of advanced manufacturing technology, which stresses the organisational aspects. As described by Zairi (1992), such a comprehensive definition has been offered by Lucas Industries according to which advanced manufacturing technology may be described as an

> integrated combination of processes, machine systems, people, organisational structures, information flows, control systems and computers whose purpose is to achieve economic product performance and internationally competitive performance. The system has defined but *progressively changing objectives* to meet, some of which can be quantified and others, such as those relating to responsiveness, flexibility and quality service, whilst being extremely important are difficult to quantify. Nevertheless, the system must have integrated controls which systematically operate it to ensure that the competitiveness objectives are continually met and which to adapt to change.
>
> (Zairi, 1992, pp.25-26. Emphasis added)

Of course, the choice between a comprehensive and less comprehensive definition always relies on a difficult choice between different levels of generality, and the question of generality is, in essence, a question of how much to include and how much to exclude. This decision faces a painfull trade-off between comprehensiveness and generality, since comprehensive definitions tend to exclude a number of real-life cases. For instance, the definition proposed by Lucas Industries leads inevitably to the conclusion that only a small number of firms on a worldwide scale qualify. However, the comprehensive Lucas definition serves the purpose of pointing out the opportunities entailed in advanced manufacturing technology and thus describes the forefront of technical progress. Furthermore, the fact that the definition is unable to capture but a small sample of firms indicates that it may serve as a *point of reference*, or analytical guide, which points to the deficiencies of many contemporary attempts at the level of the innovative firm to employ advanced manufacturing technology. Or, to state it otherwise, a comprehensive definition may, to a satisfying degree, delineate the manufacturing system of any firm *by contrast* since it allows the theorist to confront the manufacturing practice of any firm with the manufacturing best-practice of the world. Following the reasoning in chapters 2-7, the present chapter focusses on the innovative manufacturing firm and interprets the Lucas definition as a general normative prescription, or ideal-type, of technological capability. However, it will be realised that we are dealing with a normative prescription which have to be adjusted to real-life firms, reflecting that the activities of real-life firms are based on the creation of mutual theories of action within their social and cultural contexts.[125]

Chapters 8-9 applies the Lucas definition as a point of reference in four steps. *First*, section 8.2 presents some international evidence on the organisational problems entailed in the implementation of advanced manufacturing technology and outlines the nature of these problems. *Second*, section 8.3 presents some additional evidence on the initiation and implementation of process and product innovation in Danish manufacturing firms and reinterprets this evidence in terms of the flexibility-stability approach. *Third*, section 9.1 moves to a more general level by contrasting the ideal-type Fordist and Japanese management principles and discusses some new types of flexibility-stability problems which may occur in organisations that qualify according to the emerging technoeconomic paradigm described in the previous chapter. *Fourth*, section 9.2 discusses how quality management may assist in the solution of the flexibility-stability dilemma.

As argued by Jelinek & Goldhar (1983), advanced manufacturing technology is based on economies of scope as contrasted to the type of economies of scale associated with special-purpose manufacturing techniques. The important merit of advanced manufacturing technology is the absence of the trade-off between efficiency and flexibility which characterises types of technology where human skills are built into the hardware control. To put it in evolutionary terms, the trade-off between efficiency and flexibility vanishes because the technological trajectory is changed from a focus on specialised hardware to a focus on specialised software, which implies that the technological paradigm becomes focussed on programmability and reprogrammability of manufacturing techniques instead of hardware control by automated production capabilities. In consequence, economies of scope implies that economies of scale apply to cases other than large production batches.

As the outcome of this process, manufacturing becomes an information-intensive activity, not only in the sense that the usage of new production techniques is based on computerised information-processing, but also in the sense that the degree of human interaction in production increases. First, overall production planning tends to become more centralised, while production operating becomes more decentralised, and this requires that top-down information flows are matched with horisontal information flows.

economy of scope

---

125. The same line of reasoning may be applied to the macro level. Research in the field of national systems of production and innovation indicates that the application of best-practice techniques depends on the economic, social, political and cultural context in which the firm operates (Freeman, 1987; Kogut, 1991; Lundvall, 1988, 1992; Johnson & Lundvall, 1992).

decentralized *horisontal* 187
top down *information*
flow of information *flows*

As will be shown in section 9.1, this implies that learning by doing in the technical Arrow sense is replaced by learning in operational planning at the factory floor. Second, product development tends to become more costumised, which requires that intraorganisational communication linking research, design, production, marketing and sales recieves feedback from extraorganisational communication about the costumers' expectations and experience.[126] In consequence, learning by using and interacting increases in importance as a source of competitiveness, and this tendency may be reinforced to the extent that the extraorganisational selection environment exhibits increasingly shorter product life cycles as the outcome of customising and adaptability of production capabilities.

The flexibility implied by advanced manufacturing technology is, perhaps, most easily highlighted by reference to flexible manufacturing systems (FMS) and computer integrated manufacturing (CIM), which are those types of advanced manufacturing technology systems that come closest to qualify according to the Lucas definition. FMS may be described as a technical system which consists of three main parts, i.e. "computer numerically controlled general purpose machine centres with automatic tool changes, a material handling system and a mainframe computer to control the overall system" (Gupta, 1988, p.256). The concept of CIM is less technical and more organisational, since it denotes the integration of the technical and organisational aspects of productive activity from design through production and distribution in a way which utilises advanced manufacturing technology (Zairi, 1992, pp.37-41). Bessant & Buckingham (1989, p.321) define CIM as "the convergence of the various systems associated with different aspects of manufacturing around a single database and shared communications" and argue that CIM may contribute to the solution of several manufacturing problems[127], cf. the first part of figure 43. The contribution made by CIM relates to reduced lead times in design and manufacturing, accurate delivery and concomitantly falling costs associated with work-in-progress and inventories. These opportunities come about because advanced manufac-

---

126. Additionally, economies of scope implies that small and medium-sized firms becomes endowed with a number of strategic options previously only accessible by larger firms. As argued by Jelinek & Goldhar (1983), this requires a reorientation of the way in which strategic planning is carried out.

127. Their argument is based on the findings in Ferdows & Meyer (1987) which reports on the 1987 INSEAD manufacturing futures survey of European manufacturing executives. Since 1983, INSEAD has carried out an annual survey of large European manufacturing companies: "The purpose of this survey is to obtain information on the current thinking of the senior manufacturing managers on the manufacturing strategy of their business units. Over the years, the results of these surveys show the trends in manufacturing strategies, action plans, deployment of production means, and the general competitiveness of European manufacturing industry" (Meyer & Ferdows, 1991, p.211).

188   CIM   organisational

turing technology comprises a number of flexibility options in relation to production flows and profitable levels and variations of production series, as argued in the previous chapter. The second part of figure 43 decomposes these flexibility options.

However, as research in the implementation of FMS has shown, a number of obstacles may appear. *First*, advanced manufacturing technology may imply capital intensive production, if the flexibility options are not pursued: As argued by Gupta (1988), an increasing degree of capital intensity results in increased costs associated with maintenance and energy, and possibly also with labour since the increase in the expertise necessary to operate the system may lead to increased wages.

Second, the pursuit of flexibility options will only succeed to the extent that the activities of the central process of innovation (see figure 37) becomes integrated in a computerised network of communication which avoids the relay race type of innovative activity described in the previous chapter. Gupta (1988, p.261) argues that flexible manufacturing "requires a greater communication between marketing and production since short lead times and economic small batch productions enable the organization to be more responsive to the changing needs of the market" and stresses that "research and development, marketing, and production need to work together for product design". In other words, the flexibility options will be realised only to the extent that the innovative firm is able to solve performance gaps in the way described by the model in chapter 2 and further elaborated in chapters 3-7.

Third, the implementation of advanced manufacturing technology is highly sensitive to the organisational portfolio of skills and thus places a heavy burden on human resource management. This challenge relates not only to the need for new skills "such as programming, systems analysis and electronics maintenance", but also to "increased flexibility" in the deployment of skills and "a need to blend new skills with long-term 'tacit' knowledge and experience of the process involved and the materials being used and worked on" (Bessant & Buckingham, 1989, pp.326-27):

In the design area the traditional draughtsman is being replaced by a composite designer/draughtsman/CAD technician with close links into and experience of the actual manufacturing process. In the maintenance area (...) the multidisciplinary/multi-trade maintenance fitter is becoming essential to support items such as robots which involve several different technologies such as hydraulics, pneumatics, electronics and mechanical engineering.
Multiple skills are an important requirement in this connection, bringing together different engineering disciplines (hardware/software, electronics with applications, manufacturing systems engineering etc.) and different craft skills (for example, in maintenance). Further, with the decreasing importance and involvement of direct workers, those who remain need to be flexible and highly trained in first-line maintenance, diagnostics etc., whilst the

Multiple skills

increasing number of indirect support staff need to be broadly skilled and able to respond in flexible fashion to a wide variety of problems right across an integrated facility.
In essence this is a process of skill convergence to match that of technological convergence.

(Bessant & Buckingham, 1989, p.327)

Figure 43. *The contribution of CIM to manufacturing problems as seen by senior managers, and the flexibility options entailed in flexible manufacturing systems*

1. Main problems	Potential contributions offered by CIM
Producing to high quality standards	Improvements in overall quality via automated inspection and testing, better production information and more accurate control of process
High and rising overhead costs	Improvements in production information and shorter lead times, smoother flow, less need for supervision and process chasing
High and rising material costs	Reduced inventories in raw materials, work-in-progress and finished goods
Introducing new products on schedule	CAD/CAM shortens design lead time. Tighter control and flexible manufacturing smooths flow through plant and cuts door-to-door time
Poor sales forecasts	More responsive systems can react quicker to information fluctuations. Longer-term integrated systems improve forecasting
Inability to deliver on time	Smoother and more predictable flow through design and possible accuracte delivery
Long production lead times	Flexible manufacturing techniques reduce set-up times and other interruptions so that products flow smoothly and faster through plant
2. Flexibility options	Definition of flexibility options: The ability to
Product flexibility	- change to produce new products or families
Volume flexibility	- accommodate (economically) changes in volumes produced
Routing flexibility	- process parts via different routes within the plant in repsonse to breakdown or other factors
Machine flexibility	- change to make different parts within a product family
Operation flexibility	- vary the sequence of operations within the manufacturing system
Process flexibility	- produce a product family in different ways, possibly using different materials

Source: 1. Bessant & Buckingham (1989, p.322), table 1   2. Gupta (1988, p.260), table 2

190

How technological paradigm affecting management in manufactring

Fourth, the flexibility options can be pursued only to the extent that the innovative firm is able to change its organisational configuration towards the more organic principles previously described in chapter 4. Reviewing a number of studies, Bessant & Buckingham (1989) argue that the organisational structure has to be simplified across hierarchies and functions, i.e. along both vertical and horisontal lines. While automated large-scale manufacturing biased towards economies of scale implies highly specialised labour that apply skills within a structure of horisontal differentiation, advanced manufacturing technology deployed in the pursuit of economies of scope implies horisontal integration with local authority, often organised in a team-like fashion. In consequence, incentives based on rewards may become more group-oriented, and the role of supervision may change from tight control to guidance and consultancy. These differences appears in figure 44, which summarises what has been said so far about advanced manufacturing technology by comparing the ideal-type Fordist approach with the CIM approach.[128] To put it in the vocabulary of the Freeman/Perez approach described in the previous chapter, CIM, as a prescription for the implementation of advanced manufacturing technology, presents itself as a fifth Kondratieff alternative to the ideal-type Fordist organisation of the fourth Kondratieff technoeconomic paradigm.

Simplifying hierarchies & functions across verticle & horisontal

Figure 44. *Comparing the ideal-type Fordist and CIM approaches*

The Fordist approach	The CIM approach
Production of large volume, standardised products	Production of small batch, customer specified products
Dedicated production process	Flexible production process
High division between skill levels leading to tall vertical organisational structures	Increasing integration between skill levels, leading to a flatter vertical structure
Individual repetition of task emphasising horisontal differentiation	Increased integration horisontally, with semi autonomous work groups
Reward structure based on individual performance	Reward structure based on group performance
Tight supervision	Supervisor viewed as a resource

Source: Bessant & Buckingham (1989, p.329), table 3

Supervision & consultancy

128. Further elaboration of this comparison is undertaken in section 9.1.

economy of scope
horisontal integration
w/ local authority, teams

8.2. Organisational learning through creative tensions

The importance of human resource management and organic principles for the organisation of work has been stressed by Haywood & Bessant (1987) in their comparison of FMS implementation in Swedish and British manufacturing firms.[129] The management approach adopted by the Swedish firms differed from the approach of their British adversaries both in the way in which the technology systems were assessed prior to implementation and in the way in which technical and human capabilities were integrated. Contrary to their British adversaries, Swedish managers adopted a consultative approach to the initiation of FMS in order to secure that employers and employees were in agreement on the implementation strategies and shared a common view on technical and organisational innovation as a means of maintaining competitiveness rather than as a means of reducing the labour force. This consultative approach was based on a larger commitment of management towards job security, upgrading of skills and devolution of responsibilities, and in order to achieve these objectives, the time horisons of investment projects were at least three years, or almost the double of what was the average of the British sample. Furthermore, considerably more effort was put into the evaluation of and training for the systems to be implemented.

To an important extent, these differences reflected a more profound difference in industrial relations at the level of the firm regarding the attempt of management to create mutual theories of action. The Swedish firms were characterised by a greater reliance on skilled labour and low functional specialisation, and the technical and organisational innovation which FMS represented tended to reinforce these characteristics. The labour force was regarded as the most important source of competitiveness, and the procurement of capital equipment was focussed on which types of technology were manageable to the firm rather than which types of technology represented the latest development. This attitude was reflected in the fact that while the British firms devoted much more time and expenditure in computerised control functions, the Swedish firms were more oriented towards the redeployment of the existing labour force. In the Swedish firms, "automation is taking place in order to 'free up' skilled labour, but with the realisation that such systems as FMS/FMC need not, and should not, lead to the elimination of labour" (Haywood & Bessant, 1987, p.42).[130] In consequence, the Swedish firms were comparatively more

Consultative approach of FMS

---

129. The study covers thirteen Swedish firms. Haywood & Bessant (1987) are not specific on the size of the UK sample, but judged from the numbers of FMS systems which enter the comparison, the UK sample is approximately twice as large as the Swedish sample.

130. Haywood & Bessant (1987, p.3) define FMS as two "or more machine tools with a common transfer system, and under Direct Numerical Control (DNC) from a host computer", and FMC as

automation   *freeing up skilled labour*   *compare to rocketship*

successful in the implementation of FMS, measured by the costs of initial implementation and the economic benefits obtained from continued-sustained implementation, not only because they paid more attention to the reduction of inventories and work-in-progress, but also because they were better equipped to realise the quality of labour in the form of labour services through interactive learning based on mutual theories of action.[131] Table 4 compares some of the quantifiable benefits.[132]

Table 4. *Quantifiable benefits from FMS in Swedish and British firms, average figures*

Benefits, expressed in percentages	Sweden	UK
Value added, share of final product value	54	48
Labour costs, share of total production costs	18	18
Costs of work-in-progress and stocks, share of total costs	22	37
Increase of machine utilisation	74	63
Savings of manpower directly employed on production	37	58

Source: Haywood & Bessant (1987). Figures have been rounded

Based popular notion of "the factory of the future" as proverb for the firm oriented towards computer integrated manufacturing. All of the components of figure 45 fall within the range of the comprehensive Lucas definition, thus qualifying the notion of a factory of the future. As argued above, this notion implies that new principles for human resource management ought to be implemented, and in an American context[133], Moshowitz (1990,

---

two "or more machine tools with a common transfer system, but not using overall computer control via a DNC link". For the present purpose, no distinction is made between FMS and FMC.

131. Recall that section 6.3 by reference to Lazonick & Brush (1985) described learning by doing as partly social in the sense that work effort and work organisation are determined by historically specific factors, as contrasted to learning by doing in the technical Arrow sense.

132. The relative size of realised benefits should also be interpreted from the point of view that the Swedish firms had more tight, long established collaborative links, both economically and technically, with major customers, producers and sub-contractors. Thus, learning by interacting in the Lundvall (1985) sense may be assumed to be more important in the Swedish than in the UK case.

133. Moshowitz (1990) reports five site visits: "Althogether 44 individuals were interviewed - 38 managers and engineers, and six hourly workers - in four different companies" (ibid., p.623).

organisational changes must be made to realise [handwritten annotation]

p.625) has argued that although programmable automation "can continue to proceed at a fast pace without altering organization directly", the "benefits of truly integrated manufacturing cannot be fully realized without organizational changes". However, the plants visited by Moshowitz applied, by and large, a piecemal strategy to the implementation of advanced manufacturing technology, and although training programs were poorly coordinated and suffered from inadequate resources, the piecemal strategy implied that the demand for new skills could be met by the labour market. In consequence, "the provision of skills required by the factory of the future is much less of a problem than many observers make it out to be" (ibid., p.629):

piecemeal strategy [handwritten annotation]

Incremental automation, as explained above, involves a relatively small perturbation of the overall skills picture. The more ambitious, integrated approach to automation has greater impact, but here the problem of meeting the skill needs is more political than technical. The political dimension arises from questions of job security. While automation creates some new jobs for which training programs may have to be designed, it reduces labour input requirements overall. Workers are well aware of this and, although rarely hostile to technological innovation, do regard such change as a potential threat to the security of their jobs. Therefore, it should come as no surprise that workers sometimes insist that management provide training or retraining opportunities for the existing workforce as an insurance against unemployability.

meeting skill needs political [handwritten annotation]

(Moshowitz, 1990, p.629)

Figure 45. *Adaptive responses to competitive pressures*

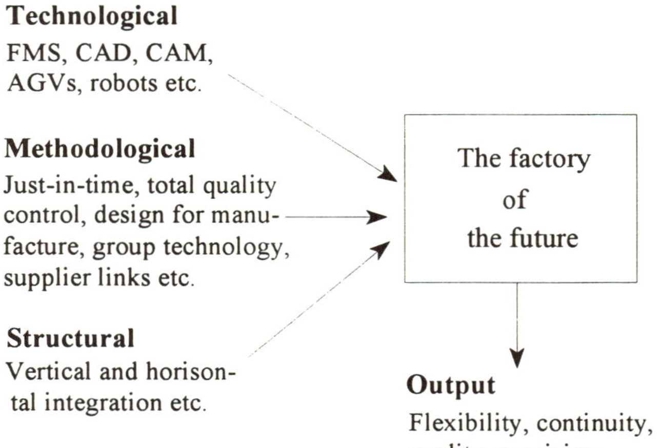

Technological
FMS, CAD, CAM,
AGVs, robots etc.

Methodological
Just-in-time, total quality
control, design for manu‑
facture, group technology,
supplier links etc.

Structural
Vertical and horison‑
tal integration etc.

*Source: Haywood & Bessant
(1987, p.6), figure 2*

The factory
of
the future

Output
Flexibility, continuity,
quality, precision,
rapid response etc.

194

not transparent

In consequence, the type of organisational changes percieved by Moshowitz (1990) relates more to industrial relations at the level of the firm than to the provision of skills. Management has to understand that the factory is a social system of production and must provide a clear committment to job security and training for new jobs. However, it was the exception rather than the case that the management communicated their investment plans to the employees, and decisions to adopt programmable automation were made primarily on the basis of technical feasibility.

While the industrial relations perspective at the level of the firm appears straight-forward from the previous discussion, and has also been demonstrated in more specific contexts such as the application of CAD/CAM (Adler, 1990), the Moshowitz conclusion on the skills issue seems less acceptable. The supply of integrative and combined skills requires that cross-disciplinary vocational training and higher education are matched with intraorganisational learning processes. Furthermore, even in the case of an incremental strategy, firms may face a shortage of labour skills, as indicated by the Danish evidence presented in section 8.3. Similar evidence has appeared in UK, USA, France and Japan where a number of surveys including the shortage of labour skills issue have been undertaken (Zairi, 1992, pp.145-54). Finally, the issue of industrial relations at the level of the firm and the provision of skills cannot be seperated, since the nature of industrial relations in the sense of manning and training policies has an important impact on the motivation of workers, and thus on the quality of labour services rendered.[134] As indicated by table 5 below, this problem seems to increase in importance as we move from advanced technical change to organisational change.[135]

incremental can still meet labor shortages

134. This relates not only to the training of subordinates, but also to the match of the skills of subordinates and superiors. A survey undertaken by the UK Amalgated Engineering Union in 1988 revealed that although new technology tended to make jobs more interesting by requiring new skills which were provided for by most firms, a significant number of the respondents thought that their superiors had inadequate skills as superiors. For instance, supervisors were assessed as weak in training skills (66%), communication skills (59%), motivation skills (59%), basic management skills (49.5%), work organisation (48%) and combined engineering skills (26%), cf. Zairi (1992, p.152, table 6.7).

135. Table 5 presents some findings of a survey on the impact of advanced manufacturing technology. The survey was conducted as interviews with "well over 4,000 managers and shop stewards at more than 2,000 workplaces in the UK" (Zairi, 1992, p.178). In sum, 15% of the workplaces and 33% of the manual workers experienced technical change over the past three years prior to the survey. The survey is reported in Daniel, W.W. (1987), *Workplace Industrial Relations and Technical Change*, London: Frances Pinter. However, the present study extracts the findings in table 5 from Zairi (1992).

Table 5. *Manual workers' reaction to different forms of change*

% of respondents, N > 4,000 Type of change	Strongly in favour	Slightly in favour	Slightly resistant	Strongly resistant
Conventional technical change	58	27	14	1
Advanced technical change	39	39	20	2
Organisational change	11	33	38	18

Source: Zairi (1992, p.179), table 7.4

The training of the labour force is an important requisite for the benefits to be obtained from advanced manufacturing technology, not only during initial implementation but also during continued-sustained implementation. For instance, as shown by Neuve Eglise (1986) in a French context, the training of personnel tended to become increasingly important in the case of CAD/CAM implementation as the firms gained experience with the new technology.[136] Furthermore, the major obstacles to a successful implementation were provided by problems which has not been considered important at the outset. Table 6 below summarises these findings.[137] As can be seen, the issues of software weaknesses, personnel training, compatibility of systems and the amount of changes increase in importance, while the more conventional problem of start-up time decreases significantly.

As argued previously, the difference between the Swedish and British approach to FMS may be explained by differences in industrial relations and in the overall management approach to the implementation of advanced manufacturing technology. To an important extent, such differences occur as the outcome of differences in the institutional set-up of national systems of production and innovation (Kogut, 1991; Lundvall, 1988, 1992; Johnson & Lundvall, 1992). For instance, Walker (1993) has partly attributed the relative weakness of the British innovation system to insufficiencies of the educational system and to poor interorganisational coordination and collaboration among the major institutional actors. Regarding the management approach, he points to the absence of industrial consensus building in the UK as compared to Germany, Japan and Sweden. In

136. The survey reported by Neuve Eglise (1986) covered 2,200 companies.

137. Similar evidence have been provided in a Danish context by Gjerding & Lundvall (1990). Section 8.3 returns to these findings.

196

Table 6. *Problems of CAD/CAM implementation in French industry*

Share of respondents who consider the problem to have a major influence on implementation, %			
Type of problem	Initial implementation	Continued-sustained implementation	Percentage change
Duration of start-up	91	28	- 69.2
Need for reorganisation	45	40	- 11.1
Software weaknesses	45	53	17.8
Training of personnel	36	51	41.7
Exchange of information	36	36	0.0
Compatibility	18	35	94.4
Amount of changes	18	36	100.0

Source: Adapted from Zairi (1992, p.50), table 3.1

contrast, Edquist & Lundvall (1993) argue that the corporatist nature of the Swedish labour market has stimulated technical change.[138] Such differences imply that although some of the implementation problems entailed in advanced manufacturing technology appear across firms and national settings, the way in which these problems are solved differs, as in the UK-Sweden comparison.

The nation-specific nature of the management approach is, perhaps, most evident in the case of a multinational company. For instance, Tyre (1991) provides some evidence on nation-specific differences in her analysis of eight plants within a leading manufacturing company located in Italy, West Germany and USA. Her analysis is based on a model of organisational response to technical process innovation which comprises the relationship between project attributes, reponse mechanisms and project outcome as outlined in figure 46. *Project attributes* are defined in terms of technical complexity, i.e. "the number, novelty, and technological sophistication of new items and improved concepts introduced (such as tooling, measurement and control systems)", systemic shifts, i.e. "the degree to which the new equipment or system introduces fundamentally changed

---

138. However, they argue that the "Swedish model" is gradually loosing its corporatist nature due to the increasing internationalisation of corporate capital which changes the power relations between the state, the trade unions and the capital owners.

manufacturing tasks or operating principles within the plant" (ibid., p.61), and the project scale measured in terms of investment. *Response mechanisms* are defined in terms of prepatory search, which is the "intensity of search and adjustment prior to the installation of the new technology" (ibid, p.62), joint search as the integrated use of external sources of knowledge, and functional overlap as defined by Galbraith (1973), cf. also chapter 4. Finally, *project outcome* is defined in terms of start-up time and operating improvement. Start-up time comprises the initial start-up and

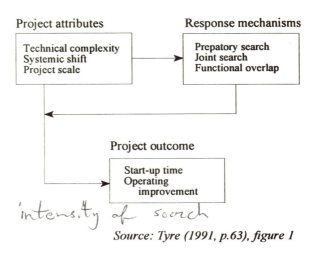

Figure 46. *Key factors in the management of technological process innovation*

Source: Tyre (1991, p.63), figure 1

the introduction periods, where the former is a subset of the latter. While the initial start-up period is the period which elapses "from delivery of the equipment until parts are being made in production mode", the introduction period is the elapsed time "from delivery until the project is considered complete" (ibid.). Operating improvement is defined in terms of "the usefulness of the technical solutions implemented, the degree to which technical objectives of the project were met, and the level of operating reliability achieved" (ibid., pp.62-63).

The regression analysis[139] of the interrelations depicted in figure 46 shows, amongst other things, that operating improvement is highly sensitive to the organisational response mechanisms applied to technical process innovation. Prepatory search, joint search and functional overlap contribute both to the speed of introduction of process innovation and to the degree of operating improvement, i.e. these types of organisational response mechanisms are highly conducive to matching successfully the stages of initiation, initial implementation and continued-sustained implementation. The European sample[140] differed considerably from the American sample in a number of respects. Regarding the relative

---

139. The regression analysis is based on a total of 48 projects, each plant comprising four to eight projects. The projects represented a spectrum of change from radical to incremental innovation.

140. The plants in West Germany and Italy are treated as an European sample, since they "were statistically indistinguishable" (Tyre, 1991, p.63).

success of process implementation, US plants experienced longer introduction times and lower levels of operating improvement. A major reason for this was that the organisational response mechanisms were invoked to a lower degree in US plants than in European plants. The US plants did not pay explicit attention to build on existing technical capacities and concepts when new equipment was purchased and did, consequently, exhibit to a larger degree than the European plants "multiple, incompatible machine control systems" (ibid., p.67). The European plants did, to a larger degree, provide intraorganisational coordination by mixing organisational roles through formal and informal integration, rotation of managerial positions and development of human resources through training and teamworking. While the Europan plants paid much attention to forming intraorganisational project groups assigned to develop the project and to integrate outside suppliers as a source of knowledge, the US plants were more familiar with buying finished technical solutions. In consequence, the European investment strategies appeared as more coherent than the investment strategies of the US plants.

Now, although the Tyre (1991) study suffers from the problem that process innovations along the radical-incremental continuum have been lumped together in the regression analysis, it serves to underline what has been argued so far, i.e. the innovative manufacturing firm will only be successful in closing performance gaps to the extent that it provides a match between technology and organisation. Advanced manufacturing technology implies, to an important extent, a technological imperative along the lines previously discussed in chapter 1, and this imperative occurs in the form of a *mutual adaptation* between production techniques and organisational configurations which is stimulated by a number of pressures for change, cf. figure 45. Leonard-Barton (1988), who employs the notion of mutual adaptation, describes the occurrence of a disequilibrium between technology and structure as a *creative tension* which, in the vocabulary of the present study, serves the double purpose of opening up and providing the stimuli for the closing of performance gaps.[141] Pressures for change occur as misalignments between the technology and "(a) technical requirements, (b) the system through which the technology is delivered to users, or (c) user organizations performance criteria" (ibid., p. 252), and the misalignments do not only imply negative performance effects, but may be

match between technology & organisation

---

141. Her point of departure is the fact that most of the innovation literature on implementation problems "focuses on either what can be done *to the technology* to adjust it to its environment or what is done to the organization *by its environment*" (Leonard-Barton, 1988, p.253). Instead, she adopts a cross-disciplinary stance and describes mutual adaptation as "the re-invention of the technology *and* the simultaneous adaptation of the organization" (ibid.). Her analysis covers twelve in-depth case studies of process and software innovation.

Implementation problems

desirable to the extent that they stimulate change. For instance, regarding the misalignment between technology and technical requirements, a

> *perfect* match between technology and the production process is not desirable. Usually technologies are introduced into processes to increase the quality of output or increase efficiency, and this improvement depends upon introducing more systematic order into the process (moving it up to higher stages of knowledge). Therefore, the transfer of new technologies into production always implies some degree of beneficial misalignment between the rigor of technology and the operation.
>
> (Leonard-Barton, 1988, p.256)

beneficial misalignment bc ↑ search

The benefits to be obtained from misalignments stem from the fact that they stimulate the search for improved solutions. Interpreted in terms of figure 12, misalignments may induce the level of aspiration to be higher than the level of achievement, thus stimulating a process of organisational learning in order to close the performance gap implied by the difference between achievements and aspirations. However, the extent to which this process of organisational learning occurs depends on the degree to which the organisation members judge (1) the activities affected by the technology in question to be significant to performance criteria, and (2) how they asses the effect of the technology on productive activities. Since technical change may have both negative and positive impacts, Leonard-Barton (1988) argues that the effect on the productive activities should not be measured as absolute gains, but as *net impact*, i.e. the ratio of positive to negative effects. Figure 47 depicts the total argument.

Figure 47. *Misalignments between technology and user performance criteria*

Four cases of misalignments between technology and performance criteria		Expected net impact on activity	
		Negative	Positive
The importance of activity to performance	High	1	2
	Low	3	4

Source: Adapted from Leonard-Barton (1988, p.258), figure 2

Interpreting figure 47 in terms of figure 16, the following conclusions emerge:[142] In case 1, the activity affected by technology is very important to the way in which performance is judged, and the expected net impact on activity is expected to be negative. Thus, case 1 may be compared to the behavioural case of poor alternatives. In case 2, importance is high and net impact is positive, and we are faced with a clear case of good alternatives. Finally, case 3, where importance is low while net impact is negative, and case 4, where the importance is low and net impact positive, may be compared to the behavioural cases of both bland and mixed alternatives.

Even though the classification presented in figure 47 is crude, as recognised by Leonard-Barton (1988), the classification is very suggestive, from two points of view. *First*, it indicates that the organisation member's evaluation of an innovation project depends on the costs and benefits that he observes or anticipates. As argued by Leonard-Barton (1988), the cost-benefit ratio may differ among individual organisation members and across the organisation.[143] This implies that the process of organisational learning may encounter the pathologies of audience and role-constrained learning described in chapter 6, and it emphasises the necessity of providing the types of visibility argued in chapter 5. *Second*, it highlights the importance of mutual adjustment of technology and structure. This is particularly important in cases 1 and 4 which exhibit a mismatch between the importance of activity to performance and the net impact on activity. It may be less important in case 2 where the mismatch seems not to occur, and in case 3 where the negative impact on activity is associated with a low importance of activity to performance.

In consequence, mutual adjustment stresses the importance of organisational learning as an interactive process. Misalignments require the construction of mutual theories of action which can only be brought about through feedback mechanisms transmitting knowledge and mutual understanding among the organisation members. This was especially the case in the Sweden-UK and European-USA comparisons reviewed above, and it implies that although learning about technical issues may take place at a high speed, learning about organisational issues may be much more erratic and troublesome. In her

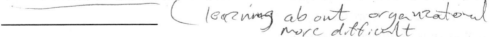
learning about organizational more difficult

---

142. Please notice that Leonard-Barton (1988) does not undertake the type of interpretation along behavioural lines which is presented here.

143. "Several of the innovations studied had a beneficial effect on the performance criteria applied to upper management, but a deleterious impact on individuals' jobs. For instance, in the case of a computer-aided engineering tool (...), the circuit designers were not interested in capturing their schematics on an electronic system. They were accustomed to handling over draft scetches to draftspeople for translation into precise schematics. Therefore the task was a relative non-significant one, and the technology impacted it negatively by requiring greater precision and detail from the engineers" (Leonard-Barton, 1988, p.259).

negative technology impact

201

study of the implementation of a manufacturing resource planning system and a purchasing control system within multiple sites of two large corporations, Leonard-Barton (1990) notices that "once new technical knowledge is acquired, it can usually be embodied in a readily transferable form" (ibid., p.165), but each

> site has some unique situational characteristics, such as parts mix and locally developed procedures, that require taylored adjustments in the technology and the organisation. Also, there is frequently inadequate continuity in job positions within the sites for someone to accumulate the organizational knowledge in one place. Moreover, much of the organizational learning is housed in people's heads rather than embodied in some readily transferable form (software, written procedures etc.), and it is not always possible to transfer the people in whose heads the knowledge resides. Finally, in many organizations, neither culture nor existing information channels support the transfer of organizational learning. In short, such transfer is not a specific part of anyone's job as the transfer of technical knowledge often is. For all these reasons, it is more difficult to identify and convey useful implementation knowledge about organizational impacts than about technical ones.
>
> (Leonard-Barton, 1990, p.169)

transfering organizational learning

In consequence, she introduces the perspective depicted in figure 48. The two curves illustrate that learning about technical issues becomes increasingly effective as experience with implementation is accumulated, while learning about organisational issues is characterised by fluctuating levels of uncertainty due to the specific characteristics of each site. Although the implementation learning curves are derived from a cross-site perspective on the level of corporation, the point of view illustrated by figure 47

Figure 48. *Implementation learning curves*

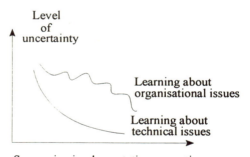

Successive implementations over time

Source: Adapted from Leonard-Barton (1990, p.166)

is applicable to the intraorganisational stage as well, as the previous discussions have indicated. It is from the analytical point of view that organisational change proves more difficult than technical innovation that the present chapter turns to the reinterpretation of some Danish evidence in terms of the flexibility-stability approach.

8.3. Some Danish evidence on the flexibility-stability dilemma

[handwritten margin note: ↑ information intensity + accountability]

Regarding the flexibility-stability dilemma, the following conclusion emerges from the empirical evidence on the initiation and implementation of advanced manufacturing presented in the previous section. Productive activities become increasingly information-intensive, both regarding the dimension of intraorganisational relationships and the dimension of user-producer relationships. Increasing information-intensity is not only created because the technological paradigm tends to focus on programmability, but also because the realisation of flexibility options require the innovative manufacturing firm to pay more attention to the integration of the activities along the chain of innovation. This implies that the creation of intraorganisational consensus and the development of skills must be based on the delegation of authority and lateral relations. Furthermore, while product life cycles tend to become shorter, the time horisons of investment projects tend to become longer, and since the rate of obsolescence of products is larger than the rate of obsolescence of processes in terms of capital equipment, the former have to be developed on the basis of more attention to user needs while the latter have to be developed on the basis of the flexibility options. Finally, the innovative manufacturing firm may experience a skill shortage which it cannot remedy intraorganisationally, and this makes the firm increasingly sensitive to the formation of skills at the labour market.

Regarding the obstacles to initiation and implementation, the evidence suggested that the types of problems which the innovative manufacturing firm encounters during implementation seem, to a small degree, to have been anticipated in advance. Furthermore, there appeared a strong positive correlation between the successful solution of implementation problems and the application of prepatory search, joint cross-organisational search and functional overlap. To some extent, the misalignments entailed in innovation serve as stimuli to change due to creative tensions which requires that the elements of the organisational repertoire are further unfreezed through organisational learning based on feedback. In consequence, the manufacturing innovative firm presents itself as an evolving system in the sense of the simple model presented in chapter 2.

In a Danish setting, further support for the conclusion presented above is provided by the studies of Högberg (1981), Gjerding & Lundvall (1990, 1992), Kristensen (1992) and Nyholm et al. (1994) which delve into the sources of innovation proposals, project championship, and stimuli and obstacles to implementation.[144] Högberg (1981) is occupied

---

144. While Högberg (1981) represents an early study in the initiation and implementation of technical innovation, the remaining studies appeared in a period where the initiation and implementation of advanced manufacturing technology was increasingly becoming an issue of political interest. Thus, the studies by Gjerding & Lundvall (1990, 1992) and Kristensen (1992)

with both product and process innovation, while Kristensen (1992) devotes his attention to product development, and the main issue in Gjerding & Lundvall (1990, 1992) and Nyholm et al. (1994) is on process innovation. In the following, the evidence presented is related to the stages of initiation and implementation.[145]

The stage of initiation

(handwritten: where do innovative proposals come from? in K-12)

Regarding the stage of innovation proposals, the important question is where the innovation proposals come from, and which types of stimuli affects the decision to innovate. Table 7 below summarises and compares some of the findings in Högberg (1981) and Kristensen (1992) regarding the sources of innovation proposals, while table 8 summarises and compares some of the findings in Högberg (1981), Gjerding & Lundvall (1990) and Nyholm et al. (1994) regarding the stimuli to innovation.

It appears from table 7 that the internal sources of ideas for innovation are closely related to management and to the development of the technical capabilities of the firm. The R&D department and key personnel were reported by 2/3 of the Kristensen (1992) sample, while approximately 1/2 pointed to top management. The external sources of ideas for innovation are closely related to the parts of the environment which are most important to competitiveness. Cooperation with customers were reported as a source for innovation ideas by 51% of the Kristensen (1992) sample, and 48% pointed to marketing & sales, i.e. the part of the organisation closest to the users of the firm's products. Compared to the Högberg (1981) sample, where top management dominates as a source of ideas for innovation, the results in Kristensen (1992) may indicate that some changes

(handwritten: ideas generated of in parts of environment closest to competetiveness.)

were supported financially by the Danish government, while the study of Nyholm et al. (1994) was undertaken by the government body itself.

145. The reader should observe that the studies are not comparable. The questionnaire undertaken by Högberg (1981) comprises 139 manufacturing firms and covers the period of 1970-80. Kristensen (1992) presents the experience of 194 manufacturing firms during 1984-88. Both samples are biased towards firms which, in a Danish context, are large. In Högberg (1981), 38% of the firms had at least 100 employees and 12% at least 500 employees. In Kristensen (1992), the corresponding figures were 73% and 35%. In comparison, the corresponding figures in the Danish population of manufacturing firms were, respectively, approximately 11% and 1% at the time of the two studies. The questionnaire undertaken by Gjerding & Lundvall (1992) covers the period of 1984-89 and comprises 337 firms of which 161 had invested in advanced manufacturing technology. Only these 161 firms enter the figures reported below. Of the 337 firms, 48.7% were engaged in mechanical engineering, as compared to approximately 2.5% of the Danish industry, and mechanical engineering accounts for 2/3 of the firms in the sample which had invested in advanced manufacturing technology. Finally, the sample in Nyholm et al. (1994) contains 515 manufacturing firms and covers the period of 1980-94. The sample is confined to seven industries and comprises only firms which can be dated back to 1984.

204

Table 7. *Sources of ideas for innovation*

% of respondents	Kristensen (1992)	Högberg (1981)
Internal sources	N = 194	N = 103
R&D department	63 '	24
Key personnel	60	--
Top management	52	49
Marketing & sales	48	22
Evaluation of products	29	--
Procurement of equipment	22	--
Production department	21	11
Design & engineering	--	6
Accountance & finance	--	1
Board	--	1
External sources		
Coperation with customers	51	
Foreign contacts	24	
Cooperation within the business group	22	
Messes and the like	21	
Only sources of ideas reported by at least 15% of the respondents in Kristensen (1992) have been included. The figures do not sum to 100% since more than one choice opportunity were given. The figures have been rounded.		

Sources: Kristensen (1992, p.22-24) and Högberg (1981, p:82)

have taken place during the time of the two studies. While the Högberg (1981) study covers the manufacturing experience of the seventies, the Kristensen (1992) study presents the experience of the eighties. In the context of the present study, one of the most important differences between these two periods is the diffusion of advanced manufacturing technology which took place at an increasing rate in the Danish economy, particularly during the mid-eighties (Kallehauge, 1992).[146] According to the arguments of the previous section, this would indicate that the intraorganisational development of the

---

146. Nyholm et al. (1994, p.59) provides some additional evidence. The diffusion of computerised communication and control seems to exhibit extremely high growth rates, as does the diffusion of computer numerically controlled equipment, automatic quality and production control, and computer aided design.

technical capabilities of the firm has become more important as a source of ideas for innovation, and this seems in accordance with the suggestions made in relation to figure 31. However, this interpretation should not be accepted without some qualification. First, it is impossible to judge the statistical significance of the difference between the two studies without further investigations. Second, the figures are not readily comparable since some overlap between the categories of top management, key personnel and other organisational functions must be expected. *stimuli types in decisionmaking*

Table 8 outlines a number of stimuli to innovation. It appears that the types of stimuli argued in the previous section play an important role in the decision of whether or not to undertake investments in advanced manufacturing technology. The flexibility options associated with the combination of increased production capacity and flexibility, reduced lead times, improved quality and production control, and the reduction of costs are reported by at least 3/4 of firms in the samples of Gjerding & Lundvall (1990) and Nyholm et al. (1994), and 72% of the Nyholm et al. (1994) sample indicate that the ability to satisfy special customer requirement, i.e. customisation, has been important. This may, to some extent, reflect an increasing awareness of the potentials of advanced manufacturing technology, since more than 80% of the firms point to increased competition and the wish to anticipate future technical advance, while more than 2/3 indicates that the decision to invest in advanced manufacturing technology has been stimulated by the production of new or improved products. Furthermore, organisational changes seem conducive to innovation, since 1/5 of the respondents in the Högberg (1981) sample reports on organisational changes and change in top management, while 55% of the respondents in the Nyholm et al. (1994) sample reports on the motivation of employees. This may indicate that investments in advanced manufacturing technology in some cases are undertaken as the outcome of deliberate human resource planning and changes of managerial roles and positions.

customization + parents as

The stage of implementation *sophisticated users*

Regarding the stage of implementation, the present section focusses firstly on implementation problems and the effects of investments in advanced manufacturing technology, and secondly on the issue of labour skills and organisational changes which were discussed at some length in the preceding section. Table 9 presents some data on the problems entailed in initial and continued-sustained implementation, while table 10 explains the importance of these problems. It appears from table 9 that more than 1/3 of the respondents pointed to problems during initial implementation associated with technical operation, lack of qualified labour and software. These types of problems relate to the operational start-up, and, as argued previously, at least technical and software problems

206

human resources & management △s

Table 8. *Stimuli to innovation*

% of respondents	Högberg (1981)	Gjerding & Lundvall (1990)	Nyholm et al. (1994)
Internal sources	N = 111	N = 161	N = 515
Decreasing profitability	42	--	--
Decreasing turnover	39	--	--
Organisational changes	20	--	--
Change in top management	19	--	--
Larger production capacity	--	84	80
Greater flexibility in production	--	80	83
Increased productivity	--	87	--
Wish to anticipate future technical advance	--	84	--
Production of new products	--	62	--
Production of new or improved products	--	--	77
Use of new materials	--	23	--
Ability to satisfy special customer requirements	--	--	72
Quality and production control	--	--	74
Networking with other firms	--	--	16
Reduction of labour costs	--	--	82
Reduction of other production costs	--	--	73
Decreased lead times	--	--	82
Increased competition	--	--	83
Motivation of employees	--	--	55
Other	29	--	12
External sources	N = 104		
New costumer demands	60		
Increasing competition	53		
New technology	47		

Only sources of ideas reported by at least 15% of the respondents in Högberg (1981) have been included. The figures do not sum to 100% since more than one choice opportunity were given. The figures have been rounded.

Sources: Högberg (1981, p.84), Gjerding & Lundvall (1990, p.429) & Nyholm et al. (1994, p.93)

might be expected in most cases. Furthermore, they tend to decrease as continued-sustained implementation is undertaken. However, some types of problems are less easy to anticipate in advance, and table 9 includes some figures which elucidates this point of view. *First*, 1/3 of the respondents experienced a lack of qualified labour which tended

stimuli & performances
gap ?

to persist during the stage of continued-sustained implementation. To some extent this might be explained by the fact that 1/4 of the respondents think that they overrated the firm's labour skills at the outset. Furthermore, problems of work organisation, which in the Gjerding & Lundvall (1990) survey include the delegation of authority, appeared in approximately 1/4 of the responding firms, and further investigations reveal that the problems of work organisation seem to persist as an obstacle to innovation (Gjerding, 1990; Gjerding & Lundvall, 1990). Tables 11-13 return to this issue.

Table 9. *Implementation problems*

% of respondents	Initial	Continued-sustained
Type of problems	N = 161	N = 160
Prolonged time of delivery	23	--
Technical problems in operation	42	15
Software	35	17
Equipment didn't meet expectations	27	--
External advice not satisfactory	35	--
Lack of qualified labour	38	26
Labour skills were overrated	26	--
Work organisation	23	--

Only problems which are reported by at least 15% of the respondents are included. Figures sum to more than 100% since more than one choice opportunity were given

Source: Gjerding & Lundvall (1990), pp.433, 439

Second, at least 1/4 of the respondents point to the fact that the equipment didn't meet the expectations and that the external advice associated with the innovation project proved to be unsatisfactory. As shown by Gjerding & Lundvall (1992, p.445), 3/4 of the firms used the supplier of the equipment as external advisor, and this might explain some of the problems described so far, since the nature of consultancy given by the supplier might be biased towards the potential benefits to be obtained from the equipment during smooth operation. The supplier may be expected to underrate the problems entailed in implementation and to be less experienced in consultancy regarding organisational problems.

The response figures given in table 9 regard the frequency of problems, and as argued some of these problems are expected to occur. However, the frequency by which a problem occurs might not be positively associated with the degree by which the success of the innovation project is sensitive to the problem in question.[147] In order to elucidate this point of view, Gjerding & Lundvall (1992) correlated the pattern of response in table 9 with the respondents' assesment of the profitability of the investment in advanced manufacturing technology. Table 10 shows the resulting rating of problems, as far as the categories in table 9 are concerned.

Table 10. *The importance of implementation problems*

Effect on profitability	Frequency of occurrence in responses	
	Very frequent	Frequent
Very serious	Lack of qualified labour	Labour skills were overrated Work organisation
Serious	Technical problems in operation	Prolonged time of delivery Equipment didn't meet expectations
Less serious	Software External advice not satisfactory	

Source: Adapted from Gjerding & Lundvall (1990), p.436

As appears from table 10, even though some problems occur very frequently according to the response pattern, the sensitivity of profitability to these problems might be less important. Instead, table 10 gives the impression that some of the problems which occur less frequently according to the response pattern are more important to profitability. These are the problems associated with labour skills and work organisation which, as argued in the previous section, are less easy to anticipate, and more difficult to solve than technical problems. Work organisation and the overrating of labour skills appeared as the most important problems during initial implemenation, and a similar rating at the stage of

---

147. In addition, it appears from Gjerding & Lundvall (1990) and Nyholm et al. (1994) that the motives for investments presented in table 8 were, to a considerable extent, fullfilled in retrospect. The two surveys included questions on whether or not the respondents regarded the motives as fullfilled after initial implementation, and the pattern of response did not differ significantly from the pattern of response associated with motives.

lack of qualified labour

continued-sustained implementation showed that the lack of qualified labour became included in the group of most important problems.[148]

The preceding section argued that a shortage of labour skills might be associated with the implementation of advanced manufacturing technology and that the manufacturing innovative firm was, consequently, becoming increasingly sensitive to the formation of skills at the labour market. However, less than 15% of the firms in the Gjerding & Lundvall (1990) survey did interact with the major institutional actors regarding the formation of skills at the labour market. Instead, 3/4 of the 83% of the respondents which provided training made use of the supplier of the new equipment, and 42% relied on intraorganisational training (total > 100% due to more than one response opportunity). Thus, the most important source of consultancy proved to be the most important source of training as well.

low exploitation

> In conclusion, the responding firms did only to a very small degree exploit the opportuniti-es for integrating external knowledge into the processes of learning which were necessary in order to solve the implementation problems that they experienced. The most important external source of knowledge was the supplier with whom the firms interacted closely. Thus, to the extent that the firms made use of extraorganisational knowledge, this type of knowledge was located in the very near environment.
>
> (Gjerding & Lundvall, 1992, p.173. My translation)

education + politics

During the nineties, the skills shortage issue has to an increasing degree entered the focus of attention of managers in manufacturing innovative firms and pervaded the political agenda for vocational training. As table 12 shows, 41% of the respondents in the Nyholm et al. (1994) survey point to the lack of qualified labour, and a vast majority of the respondents are aware of the challenges to the formation of labour skills at the levels of the firm and the labour market implied by advanced manufacturing technology. Table 11 below gives some evidence of the awareness of the skills shortage issue.

The conclusion which appears from table 11 is that generality in basic training and the availability of additional training during the time of occupation is considered important, because the technical skills of the labour force from the respondents' point of view are becoming obsolete to an increasing degree. A major reason for this seems to be that technical innovation requires additional training in multiple areas of occupation, and 3/4 of the respondents disaggree with the statement that the demand for labour skills are not greater than previously. In consequence, the skills issue is considered as an important

skills obstacle to innovation

---

148. This is, of course, hardly surprising, since a lack of qualified labour which persists through time obviously impedes the smoothness of operations and thus the profitability of investments.

affects search space & selection

obstacle to process innovation, as indicated by the figures in table 12 according to which at least 1/3 of the respondents consider the lack of skills and knowledge as an obstacle to investments in advanced manufacturing technology, alongside the need for organisational adaptation, to a degree which is comparable to more "conventional" obstacles such as the lack of finance and the difficulties of finding technical solutions which are adapted to the need of the firm.

Table 11. *Agreement with statements on labour skills*

Statement	%, N = 551
New technology requires more training for most occupational groups	92
Technical knowledge and qualifications are increasingly becoming obsolete	90
Opportunities for training during occupation are essential	89
Generality of basic training is more important than specific technical skills	81
The demands for labour skills are not greater than previously	25
Figures sum to more than 100% since more than one choice opportunity were given	

Source: Nyholm et al. (1994), p.90

Table 12. *Obstacles to investments in advanced manufacturing technology (AMT)*

% of respondents	N = 515
Previous investments have not been profitable	53
No need for further investments	51
The firm is too smal to realise the potentials	48
The use of AMT requires skills not available in the existing labour force	41
Lack of finance	37
Lack of sufficient knowledge about AMT	31
The use of AMT require organisational changes which are difficult	31
Existing solutions at the market are not satisfactory	30
The rate of obsolescence of AMT is high	19
Lack of opportunities for external finance	18
The figures sum to more than 100% since more than one choice opportunity were given	

Source: Nyholm et al. (1994), p.88

Contrary to what was argued previously about the opportunity for the firm to trade off economies of scale by economies of scope as the outcome of the implementation of advanced manufacturing technology, approximately 1/2 of the respondents consider the size of the firm as an obstacle to AMT process innovation. One important limitation faced by small firms is, of course, their lack of financial resources as compared to their larger adversaries, which implies that medium-sized and large firms might invest in advanced manufacturing technology to a greater extent than small firms. In fact, this pattern of investment behaviour appeared in Gjerding & Lundvall (1990) and is confirmed by Nyholm et al. (1994, p.65). However, the size issue seems to go further than that because only 1/3 of the respondents point to the lack of financial resources. Obviously, the avilability of financial resources is not the only reason for the difference between small and larger firms. The scale of production is an important factor as well, because a large production volume allows the firm to experiment in a piecemal fashion with advanced manufacturing technology without postponing orders and off-setting production. Furthermore, the larger number of employees in the larger firms increases the probability of the firm to possess technical and organisational skills which can be combined in trial-and-error processes. In sum, the availability of resources, time and skills imply that medium-sized and larger firms to a greater extent than small firms are able to undertake and carry through the type of adjustments in the portfolio of skills and organisational routines which are necessary during the implementation process.[149]

This argument might be validated by the evidence on organisational changes presented by Nyholm et al. (1994), cf. table 13. The previous section argued that the implementation of advanced manufacturing technology is sensitive to organisational changes which tend to create a more organic configuration, and table 13 provides some figures on the nature of the organisational changes experienced by the respondents which point to a type of organisational changes which is less frequent in small organisations. It appears that approximately four out of ten respondents reported on the formation of profit centres within the firm, i.e. the formation of subunits with local authority within budget constraints and profitability targets defined by top management, and this is a type of organisational change which is, mostly, meaningful only in larger organisations. However, the same number of respondents point to decentralisation and integration in the form of local authority of production teams, de-specialisation and just-in-time principles, which are types of organisational changes that are feasible to small organisations as well.

149. Finally, if the firm has no need for further investments, or if the firm has experienced a unsatisfactory degree of profitability in previous investments, there will be less inclination to undertake future investments in advanced manufacturing technology. This is hardly surprising.

Table 13. *The nature of organisational changes*

Type of changes	%, N = 551
Delegation of authority from management to employees	81
Smaller hierarchy of authority	75
The formation of profits centres within the firm	41
De-specialisation: The integration of tasks previously seperated	41
The formation of production teams with local authority	40
The introduction of just-in-time principles	40
Taller hierarchy of authority	13
Other	6

Source: Nyholm et al. (1994), p.99

In conclusion, the small/large size issue must remain unsolved in the present context which is based on the Nyholm et al. (1994) survey. Instead, the discussion indicates that the type of organisational changes which are undertaken during the implementation of advanced manufacturing technology may vary according to firm size, the formation of subunits being an option to primarily larger organisations. Whichever the case, the average conclusion to be derived from table 13 is that most of the respondents point to changes associated with the lines of authority and the match between horisontal and vertical communication and coordination. In consequence, 3/4 of the respondents experienced that the hierarchy of mangement became smaller, and 81% point to the delegation of authority. In contrast, only 13% of the respondents thought that the hierarchy of management became taller.

communication

8.4. Looking ahead: Some final remarks on skills and organisational principles

The discussion in sections 8.2-8.3 indicates that the mutual adjustment implied by advanced manufacturing technology have to be based on a match between the formation of skills and the creation of organic forms of intraorganisational coordination. Freeman (1992b) argues that the type of technical change associated with the fifth Kondratieff implies a strong growth in technical, scientific and related professional occupations, and managerial and administrative occupations on the basis of an increased demand for the

skills along these lines of occupations.[150] It seems reasonable to hypothesise that the proliferation of these types of skills is likely to create an intraorganisational demand for discretionary power at lower levels of the hierarchy of coordination, and concomitantly organisational tensions which can only be solved through organic forms of intraorganisational coordination.[151]

However, Freeman (1992b, p.184) argues that this trend may be changed as information and communication technology "becomes a more familiar and possibly more standardised set of technologies". According to Sundqvist et al. (1988, p.70), information technology has affected the organisation of work in a "variety of ways", and although they discharge the possibility that information technology may promote Taylorist and low-skill organisation on efficiency grounds, they recognise, at the same time, that information technology may promote a differentiation of skill-using strategies: -

> When equipment and materials are valuable, higher skill in the workforce need produce only a moderate saving in down time to offset the training and recruitment cost involved in those for less skilled labour.
> In any event, equipment embodying new technologies reduces capital requirements much more frequently than it raises them. (...) In such cases, although the propensity to dedicated modes of utilisation is reduced, so too are the incentives to develop a high technical level in the workforce. Skill-utilising strategies are thus less likely in sectors where capital requirements remain low or are reduced by new technology, such as textiles, clothing and footwear.

(Sundqvist et al., 1988, p.71)

At the level of the firm, the differentiation of skill-utilisation becomes even more clear in the case of advanced manufacturing technologies that are approaching a relatively high degree of maturity, i.e. standardisation, such as computer numerically control. In the case

---

150. Freeman (1992b, p.183) points to labour market shortages throughout the OECD economies as evidence on an increased demand for scientific, technical and communication skills. In a Danish context, Banke & Clematide (1989) indicate the same.

151. This might, especially, be the case in a Scandinavian and European context. For instance, Freeman (1992b) argues that "more skill-intensive solutions are often found in Scandinavia and Germany than in the United Kingdom or the United States" (ibid., p.183). As argued previously, this may partly be explained by the nation-specific traditions of industrial relations. In fact, it has less to do with the proliferation of advanced manufacturing technology which according to Edquist & Jacobsson (1988) has diffused at a broader range and higher speed in the Anglo-Saxian countries than in Northern Europe. Industri- og Handelsstyrelsen (1988) provides some Danish evidence which are more detailed than the surveys by Gjerding & Lundvall (1990) and Nyholm et al. (1994), and comparable surveys in a British and Swedish context are presented by Northcott & Walling (1988) and Statens Industriverk (1988), respectively.

of CNC machinery, Banke & Clematide (1989, pp.39-46) report on seven manufacturing firms which in my eyes may be grouped into three categories according to the division of labour. Banke & Clematide (1989) divide CNC-tasks into nine groups, i.e. (1) programming, (2) testing of programs, (3) set-up, (4) preparation of tools, (5) surveillance, (6) admittance and removal of materials and blanks, and operation, (7) control of operation, (8) control of finished and semi-finished products, and (9) maintenance and repair. While two of the firms integrated (2)-(8), two firms integrated (2)-(4) and (6)-(7), and the remaining three firms exhibited a larger diversity, however with the integration of (5)-(7). Thus, although some similarities are present, the degrees of freedom are numerous, as described by Banke & Clematide (1989, pp.47-52).

Even if the degrees of freedom in CNC operation may, eventually, lead to an integration of tasks at the shopfloor, numerical control may contribute to a larger *vertical* division of labour. Cavestro (1986) argues that numerical control may contribute to either seperate or bringing closer together the planning and the execution of work. A seperation of planning and execution seems to appear in the context of small and medium-sized French firms, where there emerges "a growing dichotomy between the shopfloor and production and engineering activities" (ibid., p.129), and where automation "rapidly generates a specialisation of tasks which tends to dissociate the planning of production runs from programming" (ibid., p.128). This dichotomy appears to be associated with the type of technology in question: In the case of CNC, production engineering takes place in close contact to the workshop; in the case of DNC, a more hierarchial organisation of technical work seems to apply.

In consequence, one may hypothesise that as integration of the various types of advanced manufacturing technology becomes more standardised, a larger variety of organisational principles than implied by sections 8.1-8.2 will emerge. At the shopfloor level, organicity may be larger than at the vertical level of the organisation, and at the level of project groups or task forces, the degree of organicity may become quite large[152]. Thus, one might imagine a type of organisation where initial implementation assumes a high degree of complexity and a low degree of formalisation and centralisation, and where continued-sustained implementation is characterised by the same properties as at the shopfloor while formalisation and centralisation in relation to planning activities increases relatively to the shopfloor in the vertical dimension. Depending on the types of technology in question, the same may apply to engineering.

However, even though the relative differentiation of skill-utilising strategies may become an actual trend, the degree of complexity may increase since the skills content of

---

152. Christensen, J. (1992) reports on some interesting cases in this respect.

formal & control

215

human capital presumably continues to increase. Thus, firms employing advanced manufacturing technology may become more receptive to innovation, but may, at the same time, increasingly encounter the types of flexibility-stability problems described in chapters 2 and 5. In sum, the nature of the flexibility-stability dilemma faced by the manufacturing innovative firm may be characterised by an increasing emphasis on the dilemma of complexity relative to the dilemma of formalisation and centralisation.

If this future trend actually comes into existence, the manufacturing innovative firm faces an intricate problem of maintaining organisational coherence. A larger degree of receptiveness to innovation may increase the amount of opportunistic surveillance relative to problemistic search and thus induce an inclination of search behaviour towards distant search and global rationality.[153] From the neo-behavioural perspective described in chapter 6, one may argue that since opportunistic surveillance stimulates the proactive perception of performance gaps, the intraorganisational theories of action may become embedded with relatively more solutions looking for a problem than problems looking for a solution. A situation like this imposes on the organisational configuration a number of creative tensions, because it adds a new dimension of diversity to the feedback and lateral processes by which the mutual theories of action are created and maintained. Consequently, from a behavioural point of view, the fifth Kondratieff techno-economic challenge to contemporary firms may promote stress or slack innovation relative to innovation occasioned by necessity or programmed action.[154]

This proposition may, to some extent, be elucidated by the comparison of the ideal-type Japanese and Fordist principles of work organisation undertaken in section 9.1, which takes as its point of departure the contrast between the CIM and the Fordist approach previously described in figure 44.

---

153. Please recall figures 14 and 25.
154. Please recal figure 13.

Chapter 9
New Management Principles and Quality Control

9.1. Flexibility-stability dilemmas in an ideal-type Japanese setting

The contrast between ideal-type Japanese and Fordist principles of production and work organisation has often been invoked by those who try to explain the instances at which Japanese firms have superseeded their Western adversaries. A number of interesting stories have been told. For instance, take the sad tale of Xerox that invented the modern copy machine but at the end of the 1970s had lost its competitive edge *vis-á-vis* Japanese competitors able to produce at 50% of the manufacturing costs, needing only half the staff in research and development with half the time of product development (Dertouzos et al., 1989). Similar cases may be found in consumer electronics, semiconductors and computers, and one may furthermore be inclined to tell the story of Honda that set up and expanded production facilities in the US so successfully that Honda became the sixth largest automobile manufacturer in USA within three years (Insley, 1989). The loss of the competitive edge in American industry as compared to its Japanese competitors has often been explained by the techno-economic shift between the fourth and fifth Kondratieff in the sense that those firms, which were the most successful during the Fordist era, have become the victims of the very large-scale, mass-consumptionist principles that constituted their initial success. However, this is only one part of the story, and it does certainly not explain the success of the Japanese management system at the very hart of mass production, i.e. the automobile industry.

Gjerding (1992) argues that the existence of this phenomenon rely on national differences in how organisational learning is contingent upon the principles of work organisation which reflect the differences between *myopic* and *dynamic* systems (Pavitt & Patel, 1988) of innovation. According to this analytical perspective, fallacies of the Fordist management system are rooted in the following characteristics:[155] A focus on short term financial performance[156]; the lack of a technically and organisationally skilled

flaws in Fordism

155. These characteristics are derived from a number of sources such as Leibenstein (1987), Freeman (1987, 1988), Urabe (1988), Dertouzos et al. (1989), Dunning (1990), Whittaker (1990, 1990a), Lincoln (1990), Roos (1991), Shimada (1991) and Aoki (1984, 1990, 1991).

156. The focus on short run financial performance is, especially, associated with national systems of innovation where the financial system is based on market operations rather than lender-borrower relationships, i.e. a credit-based financial system. Christensen, J.L. (1992) provides an interesting discussion of the implications for technical change.

NSI

how education?

workforce; the treatment of the workforce as a commodity rather than as a long-term investment in competitive capabilities; the neglect of the importance of close cooperation between design and manufacture; and a resistance to delegating authority downstream to the shopfloor level. The present section, which is based on the argument presented in Gjerding (1992), elaborates on this point of view in terms of five analytical concepts: The overall management approach; logistic principles; the human resource perspective; the degree of intrafirm specialisation; and the decision making process. Figure 49 summarises the following discussion.

Figure 49. *The ideal-type Japanese management system*

The overall management approach	Integration of - stages of production - human and physical capital - development of strategy and human capital
The kanban philosophy	Demand-pull production flows Just-in-time Decentralised planning and control

Humanware	Human controls software, not the opposite
	Machinery: Means, not goal
	Continous development of human capital
	   - job rotation
	   - retraining
	   - additional training and education
	The process of work: A process of learning
	Employees: The key to change

Intrafirm specialisation	Task specialisation, however within broad job categories

Source: Adapted from Gjerding (1992a), p.18

The overall management approach
According to Shimada (1991), the Japanese management model may be described as an *integrative* model as opposed to the *confrontational* model characterised by the Fordist principles of production. While the confrontational model "is designed to minimise the influence of human variability upon the performance of the production system", the integrative approach emphasises the interaction between the social and technical systems,

and accepts that "human variability affects significantly the performance of the production system" and that the "performance of the system will depend critically on human actors and vice versa" (ibid., p.460). Similarly, Itami (1988) applies the notion of *peoplism*, i.e. a management philosophy according to which the

> employees (or at least long-time employees) are regarded as the de facto "owners" of the firm, because they supply the most precious resource - human resources.
>
> (Itami, 1988, p.28)

Two important properties may be attributed to the ideal-type Japanese management system: *First,* income shares tend to become equalised, not only horisontally, but also vertically (Lincoln, 1990). This is the outcome of a sharing system, which is parallelled by the distribution of intraorganisational power since even shopfloor workers, contrary to the Fordist practices, are equipped with some discretion over technology and the production process (Aoki, 1990). Thus, the relationships between power and remuneration, and power and vertical organisational roles, are weak. *Second,* following Freeman (1987, 1988), the Japanese success may partly be attributed to dynamic learning economies based on *reverse engineering.* Reverse engineering implies that the entire production process is percieved as an integrated system based on the use of the factory as a laboratory characterised by the integration of tasks. Product and process innovation are regarded as an integrated activity, and incremental innovation combined with an emphasis on quality control provide a dynamic contribution to product quality and productivity. In consequence, reverse engineering spurs competitiveness through improvements of the product mix and production efficiency.

While the nature of the sharing system must be assumed to mitigate the flexibility-stability dilemmas associated with stratification and job satisfaction, the integrative model may be faced with a dilemma of complexity similar to the one described in the concluding remarks of section 8.4. The integrative model is characterised by an appreciation of human resources which includes fluid boundaries between tasks, delegation of authority and an emphasis on the ability and propensity of the workforce to engage in learning. Thus, the number of attention centres which are bound to interact is large, and similarly the reliance on coordination through communication. In consequence, the economic performance of the integrative model depends on smooth and open-minded feedback processes along both vertical and horisontal lines in a way that stimulates local discretion without creating a large differentiation of subgoals. This is an intricate balance which is highly sensitive to the creation of mutual theories of action.

Logistic principles: Kanban

The confrontational model implies that the production flow is coordinated through a push system which relies on central production planning with precise and specific instructions running down the production line. Contrary, the integrative model is based on a pull system where the output level of an operational unit is determined by the level of demand that occurs downstream (Shimada, 1991). The pull system is signified by just-in-time principles, which according to Urabe (1988) is the strategic core of the *kanban* system. Kanban aims at stockless production, high quality and productivity rates, and contributes to make small-lot production economically feasible to a larger extent than may be implied by the flexibility options presented in chapter 8. Following Aoki (1990), the kanban system may be described as the combination of (1) central tentative production planning that provides a general guide-line for a specific period of time, and (2) local discretion over technology according to interunit demand stimuli within the framework of the general guide-line. In consequence, the functioning of the system is sensitive to the communication of intraorganisational demand, product defects and machining problems. It forces the employee to respond to changes in intraorganisational demand, exercise control over local emergencies and provide self-management and self-inspection. Furthermore, the employee is encouraged through experience to make minor modifications to the production process in order to secure smooth operation.

The kanban system enhances the receptiveness of the organisation to innovation in a number of respects. As argued by Roos (1991), kanban is designed to reveal problems rather than to override them in a Taylorist fashion, and the single most important feature of the system is the flow of horisontal information in order to detect and prevent obstacles to local activity programmes.[157] This may have at least two important implications: *First*, the search for efficient solutions is enhanced. This implies that the levels of aspiration is continously pushed upwards above the level of performance. *Second*, the accumulation of experience at the horisontal level, especially through learning by doing self-inspection and self-management, induces the employee to percieve causes of local emergencies and anticipate the need for future changes of local activity programmes. In consequence, problemistic search and opportunistic surveillance may occur more frequently in an ideal-type Japanese setting than in an ideal-type Fordist setting. While solutions to emerging problems might be obtained through consultation with experienced peers, i.e. scanning the intraorganisational environment for nearly-finished solutions, solutions to anticipated

---

157. In order to improve product quality, eliminate waste (*muda*) and stimulate continous improvement of the production process (*kaizen*).

problems are harder to get at, and thus not only reproductive, but also productive search are likely to occur more frequently.

The human resource perspective

The human resource approach of the Japanese management model as opposed to the Fordist model may be exemplified by the application of CNC, as evidenced in the Anglo-Japanese comparison by Whittaker (1990, 1990a) who focusses on the human-machine interface.[158] Whittaker (1990) distinguishes between organisation-oriented and market-oriented industrial relations and argue that CNC operators experience more training and a broader range of tasks in cases of organisation-oriented than market-oriented relations. This seems to hold true in the Anglo-Japanese comparison, where the main impression is that British workers are hired to do specific tasks, while Japanese workers are hired to a specific *range* of tasks.

This difference is reflected in the approaches of British and Japanese managers to the use of CNC. The British managers view CNC as a machine tool with a computer attached to it and emphasise the ability of the operators to secure smooth machining operation, while the Japanese managers view CNC as a computer with a machine tool attached to it and thus as a device that functions properly provided correct programming is undertaken (Whittaker, 1990a). Consequently, the Japanese managers emphasise the ability of the employee to learn computer programming, while the British managers emphasise machining experience. As a result, programming and operating tasks tend to be intertwined in the Japanese case, thus leading to broad job classifications, and seperated in the British case where narrow job classifications are maintained.

The difference between these two approaches may be described as the difference between a technical and a craft approach, as depicted in figure 50. The technical approach of the Japanese managers, as opposed to the Fordist-oriented craft approach, is associated with a high degree of continous on-the-job training and retraining (Freeman, 1987), and the acquisition of diagnostic skills through participation in quality control, notably in the well-known quality circles (Whittaker, 1990a). Employees are regarded as important "agents of change" (ibid.), and contrary to the Fordist practice where on-the-job training typically takes the form of a short instruction within a narrow task range, the technical on-the-job training focusses on learning through experience where newly-hired employees often are supported by full-time teachers and team-leaders (Shimada, 1991).

---

158. Whittaker (1990, 1990a) reports on eighteen case studies, matching nine British and nine Japanese firms in pairs approximately equal in size, product technology and batch size.

Figure 50. *The technical and craft approaches to CNC*

	The technical approach	The craft approach
Unmanned operation	Seen as an attraction of CNC and actually carried out	Not seen as an attraction, and an operator has to be present
Experience on manual machines	Operators have often little or no experience on manual machines	Operators have considerable experience on manual machines and often a craft background
CNC-specific training	Operators chosen for CNC are given relatively less CNC-specific training and are sometimes required to master it themselves	Operators chosen for CNC are given relatively more CNC-specific training
Multimachine operation	Seen as an attraction of CNC, and less experienced operators can operate on more than one machine	Not seen as an attraction, and experienced operators has to concentrate on one machine
Programming	Specialist programmers are not required to have machining experience, and almost none have a shopfloor background	Specialist programmers are required to have considerable machining experience, and almost all have a shopfloor background

Source: Whittaker (1990), p.157

The technical approach is conducive to the development of company-specific skills through the type of learning by doing described above, which according to Urabe (1988) provide an important stimulus to incremental innovation through continous improvement associated with minor adjustments of process technology and proposals for minor adjustments of product technology. The incremental improvements of hardware implies that the production equipment is no longer subject to automatic decay and depreciation but rather can be an asset, the capacity of which may improve and appreciate over time as a result of the interaction with human resources (Shimada, 1991, p.462).[159] In consequence, the type of learning by doing which takes place in the ideal-type Japanese setting is more dynamic than learning by doing in the Arrow sense, because it is based on a match of interaction between peers and interaction between humans and machines. The organisation of intrafirm education and training is based on the exchange of information and experience between the organisation members in a way which is likely to stimulate

---

159. This process is often designated by the notion "giving wisdom to the machines".

both problemistic search and opportunistic surveillance to a larger degree than in a craft setting. However, as argued in relation to the overall management approach, this may entail the problem of differentiation of subgoals and thus the dilemma of complexity.

The degree of intrafirm specialisation

The degree of intrafirm specialisation relates to the vertical division of labour between hierarchial levels and the horisontal division of labour between job functions. In each dimension, standardisation of tasks takes place. Scientific management and the Weberian-type bureaucracy are prototypes within the Fordist model where standardisation is achieved to the smallest economically feasible unit. However, in an ideal-type Japanese setting the unit of standardisation is "not the job function of an individual worker, but rather a group of job functions performed by the team of workers" (Roos, 1991, p.107). Thus, the technical approach implies that the smallest economically feasible unit is quite different from that of the Fordist model. Employees become organised in teams associated with a cluster of interconnected jobs among which the members rotate, and the borders between jobs become concomitantly fluid and ambiguous (Aoki, 1990). In consequence, where the number of job classifications may be very large in an ideal-type Fordist setting, the job classification system in the ideal-type Japanese setting confines itself to a few, broad job classes (Shimada, 1991).

This phenomenon has important implications for the ability of the organisation to avoid the problems entailed in the dilemmas of complexity and stratification. *First*, a broad job classification system, which is based on teams and the rotation of jobs, avoids the development of "property rights" in the job because the organisation member find it difficult to identify himself with a special assignment (Leibenstein, 1987). *Second*, the system implies that the workforce becomes multi-skilled and flexible and thus easier to redeploy "as circumstances change" (Lincoln, 1990, p.11). Consequently, the organisation is able to trade off lower economies of specialisation with higher economies of multi-functionality and attendance to local emergencies. This is a kind of dynamic efficiency which is based on "collective learning by workers and encouraging semi-autonomous problem-solving and adaptation to local schocks by the versatility of workers at the shopfloor level" (Aoki, 1990, p.277). The versatility at the shopfloor level may increase the frequency of solutions looking for a problem, but in a way that may prevent audience and role-constrained learning and stimulate the amount of feedback.

The decision making process

An important part of the flexibility-stability dilemma in a Fordist setting is the problems associated with stratification and job satisfaction which induce the organisation members to feed back only positive information on their job performance. This phenomenon is further enhanced if the degree of formalisation and centralisation is high. However, these types of problems may not occur in an organisational setting based on a match between local discretion and collective learning, which implies that the stock of knowledge is increased as part of the creation of mutual theories of action. However, the ability to trade off lower economies of specialisation may be hampered unless the incentive system is in accordance with the creation of mutual theories of action.

The logistic principles and the human resource approach described earlier place a heavy burden on the incentive system, because they require a high degree of motivation on behalf of the organisation member to adapt to changing circumstances by exercising local discretion. In the present context, the following three components of the incentive system is considered: Remuneration, carreer ladders and stimuli to individual creativity. Shimada (1991) argues that the motivation to adapt to changing circumstances is affected by wages only to a limited extent while promotion and job rotation are much more important:

> Reward systems such as this doubtless not only have a significant impact on worker motivation; they also stimulate the adaptability of workers, while basic wages are not affected much by the transfers or rotation among different job assignments. To the extent that scheduled transfers are used as instruments for workers' career formation, the promotional rewards may well foster greater adaptability of workers.
>
> (Shimada, 1991, p.466)

The promotion and job rotation components of the incentive system in the sense described by Shimada (1991) tend to stimulate the processes of feedback and thus organisational learning, because they provide a basis for overcoming the conceptual and linguistic barriers to intraorganisational communication. In the ideal-type Japanese setting, this property is further enhanced by an institutional set-up which implies that status and job responsibility are decoupled (Lincoln, 1990), and this stimulates the equalisation of blue collar, white collar and management employees in a way that according to Urabe (1988, p.13) contributes "to the elimination of status barriers in communication". An important part of the institutional set-up is the *ringi* practice, i.e. the circulation of innovation proposals briefly sketched out on a piece of paper that may originate at any level of the hierarchy; the *nemawashi* practice, i.e. informal consultation across formal hierarchial levels; quality circles; and numerous other kinds of formal and informal meetings and communications. This set-up provides a number of instruments by which mutual theories of action can be created, and especially the ringi practice, which is an intriguingly simple way of communicating ideas, may stimulate the desire of organisation members to engage in prepartory search through opportunistic surveillance.

In sum, the ideal-type Japanese management model seems extremely efficient in order to solve the flexibility-stability dilemmas originally described in chapter 2 and further elaborated in chapters 3-6. Thus, one might hypothèsise that in order to realise the flexibility options associated with advanced manufacturing technology, Japanese-style organisational principles should be applied after some adjustment according to the specific national setting of the organisation. However, a number of problems related to the dilemmas of complexity, formalisation and centralisation, and stratification and job satisfaction may occur:

(1) Instead of the problems of code scheme barriers and insufficient creation of mutual theories of action, the Japanese-style organisation may experience an extensive burden on the incentive system, because the rate of change of activity programs and thus the demands on the adaptability of the organisation members may become too high. The occurrence of this type of dilemma depends on the coordination of local discretion.

(2) A high degree of routinisation and restricted information flows may be substituted by an insufficient degree of routinisation and an information overload. In consequence, the demands on the organisation member's span of attention outgrow his capability, and the demands on the creation of mutual theories of action outgrow the organisation's ability to act accordingly since the differentiation of subgoals tends to proliferate.

(3) While the Fordist problems related to stratification and job satisfaction may be absent, the decoupling of status and responsibility may spur the rate of innovation proposals to such an extent that the organisation is overfloated by solutions looking for problems. In this case, a number of organisation members will become dissatisfied with the rate of change of activity programmes and experience the state of affairs as audience learning. In fact, Lincoln (1990) argues that the efficiency of the famous quality circles is deteriorating because an increasing number of Japanese Q-circle participants experience the Q-activities as an extra burden on their work effort that yields too few results in actual implementation.

9.2. The opportunities provided by quality management

This is a provocative observation, since the application of quality management is often referred to as an important source of the relative competitiveness of the Japanese management system. While an ideal-type Fordist setting emphasises quality *control* in terms of conformance to requirement, the appropriate notion in an ideal-type Japanese setting may be quality *management* implying an overall focus on cooperation within the firm. Actually, Japanese managers and theorists prefer the concept of *company-wide quality control* (CWQC) that according to the Japanese Industrial Standard Z8101-1981 may be described in the following way:

> ...implementing quality control effectively necessitates the cooperation of all people in the company, involving top management, managers, supervisors, and workers in all areas of corporate activities such as market research, research and development, product planning, design, preparations for production, purchasing, vendor management, manufacturing, inspection, sales and after-services, as well as financial control, personnel administration, and training and education. Quality control carried out in this fashion is called company-wide quality control.
>
> (Sullivan, 1988, p.11)

Figure 51. *The ISO definition of a quality management system*

Quality management	The aspect of the overall management function that determines and implements the quality policy
Quality policy	The overall quality intentions and direction of an organisation as regards quality, as formally expressed by top management
Quality system	The organisation structure, responsibilities, procedures, processes and resources for implementing quality management
Quality assurance	All those planned and systematic actions necessary to provide adequate confidence that a product or service will satisfy given requirements for quality
Quality control	The operational techniques and activities that are used to fulfil requirements for quality
Quality plan	A document setting out specific practices, resources and sequence of activities relevant to a particular product, service, contract or project

Source: DS (1988, pp.8-9) and Møltoft et al. (1991, p.17)

The CWQC concept is in accordance with the concept of quality management applied by the ISO 8402 standard, cf. figure 51. The ISO 8402 quality vocabulary was originally proposed by the International Standards Organisation (ISO) in order to present a definition of quality management that could resolve the ambiguity which pervaded the international literature and practice within the field. At the outset, the vocabulary defines quality management as "the totality of functions involved in the determination and achievement of quality" (DS, 1988, p.3) which in some parts of the international research has been identified as quality assurance. However, in the ISO 8402 the term "quality assurance" refers exclusively to the case of a negotiated relationship between a user and a supplier, where the user poses some requirements on the quality management of the supplier. The method of definition employed in the ISO vocabulary is hierarchial, as depicted in figure 51, and *quality* is defined as the "totality of features and characteristics of a product or service that bear on its ability to satisfy stated or implied needs" (DS, 1988, p.6).

Studies in the field of comparative quality management often mention that although the theoretical and practical evolution of quality management originated in an American context, where the concepts of quality costs and total quality control came to be associated with the work during the 1950s by Joseph Duran and Armand Feigenbaum, the broader approach to the management of quality came not to be associated with the American, but the Japanese management model. The fact that Joseph Duran and W. Edwards Deming spent quite some time in Japan during the 1950-60s teaching the management of quality to Japanese managers is often employed as an explanatory factor in studies of the Japanese lead in the field of quality. Apparantly, the Japanese managers were more receptive to the idea of quality control, and Deming (1982, pp.99-110) argues that this may be attributed to five important forces: The motivation and capability of Japanese statisticians; the establishment of the union of Japanese science and engineering (JUSE); the formation of a training scheme in statistical methods that eventually build statistical knowledge into the engineering education; the inclination of Japanese managers to consciously interpret the issue of quality as everyone's job; and the invention of quality circles. While the stimuli to most of these forces came from the US, the quality circle is a genuine Japanese conception.

Quality management represents a strategy for potentially solving some of the flexibility-stability dilemmas described throughout this study, especially if quality management is interpreted in terms of the *Deming philosophy*. In fact, the ideal-type Japanese management system described in the previous section, cf. figure 49, is highly conducive to the application of the Deming philosophy according to which quality is achieved

> by the coalescence of two forces: total teamwork and the 'scientific approach'. The scientific approach requires understanding of the nature of variation, particularly its division into controlled and uncontrolled variation due to management-controllable common and worker-controllable special features. It is not only by management and workers correctly diagnosing the most important sources of variation, and then reducing or even eliminating them, that quality (reliability, consistency, predictability, dependability) can be improved. (Edge, 1990, pp.397-99)

Deming (1986) argues that the quest for excellence in the management of quality can only be realised through the combined committment of management and employees on the promotion of quality. This requires an increased emphasis on (1) the elimination of the need for inspection by building quality into the products through process quality; (2) the institution of on-the-job training, education and self-improvement; and (3) the removal of organisational barriers between departments supplemented by the affinity of employee authority to employee responsibilities, as opposed to the type of coordination entailed in management by objectives.[160]

These organisational prescriptions, which Edge (1990a) denotes by the concept of "the factory as a people-driven system"[161], constitute the essence of Deming's now famous fourteen points for management, cf. figure 52, which describe the organisation as a coherent team where the rationalities at the different levels of the hierarchy are aligned by a system of open horisontal and vertical communication, i.e. a system of communication and information processing where problems are attended to through illumination rather than concealment. Of course, the notion that "quality is everyone's job" does not imply that the lines of authority are blurred or subjected to the principles of political anarchy. What the Deming philosophy proposes is that while a uniform direction of purpose is the responsibility of management, the evolution of that purpose should be instituted in the form of lower-levels discretion and bottom-up feedback mechanisms.

The feedback of information and perceptions in relation to quality depends, of course, on the type of lateral processes employed by the organisation. As emphasised by Deming (1982) and Gitlow & Gitlow (1987), the structuring of top management for quality involves that all levels of the organisation are structured for quality, preferably in a team-like fashion. At present, quality committees and quality circles are the most outstanding ways of stimulating the necessary lateral processes and provide direct contact between managers, the creation of liason roles, integrating roles and managerial linking mechanisms, and a flow of horisontal and vertical communication that facilitates learning.

---

160. Management by objectives implies that the superior and the subordinate reach an agreement on the goals which the subordinate is to achieve. When the agreement is reached, an appropriate organisational role is negotiated and the subordinate is provided with a specific area of authority and responsibility. However, if it turns out that this area is insufficient to reach the goals agreed upon, negotiation has to be resumed. In consequence, processes of feedback and improvement through changes in lower-levels discretionary behaviour are impeded.

161. Similarly, Joiner & Scholtes (1988) argue against "the negative aspect of management by results" and in favour of "quality leadership".

Figure 52. *Deming's fourteen points for management*

The fourteen points	Interpretation
Create constancy of purpose towards improvement	Do not let the problems of today dominate the problems of tomorrow. Allocate resources for long-term planning, research, education, maintenance and continual improvement of design
Adopt the new philosophy	Recognise that the improvement of quality is the key to productivity growth and job security
Cease dependence on mass inspection	Inspect all or inspect none. If inspection is too costly, don't do it. Improve the process and build in quality instead
Change the philosophy of purchasing	Do not purchase on the basis of a price tag alone. Require, instead, statistical evidence that quality is built in
Find problems and improve the system	Institute a PCDA schedule: Plan for action; Do the act; Check the results; Act by making the necessary subsequent changes
Institute modern methods of training	Workers must learn to achieve statistical control and understand quality goals. Training must be evaluated and continued
Institute modern methods of supervision	Align responsibility and task and avoid performance appraisal. Supervision should advance feedback, training and pride
Drive out fear	Provide psychological and physical job security, encourage horisontal communication and delegate discretionary power
Break down barriers between departments	Stimulate vertical and horisontal verbal communication. Train for teamwork and form teams with discretionary power
Eliminate goals without means	Replace numerical goals for the workforce by the combination of mission statements and statistical methods
Eliminate management by numbers	Replace numerical quotas for the workforce by statistical methods, training, cooperation and road maps to quality
Eliminate the barriers to job satisfaction	Involve employees at all levels, and train and job rotate in order to provide understanding of the extended production process
Institute education and retraining	Provide training in relation to the overall goals, then statistical techniques. Secure training when technical and organisational change are necessary
Structure top management for quality	Institute quality committees and quality circles. Provide statistical subunits, assistance or capability for each

Source: Deming (1982, pp.16-50) and Gitlow & Gitlow (1987, pp.13-201)

However, although the Japanese experiences with team-oriented organisational action in the field of quality seem promising, it has been difficult to supplant this idea to other types of national contexts. Reviewing a number of studies, Wollfberg & Saabye (1988) point to three main causes for the defect of quality circles:

(1) A lack of motivation and support from top management. Q-circles are percieved more like devices for human resource development than as an important part of a quality management system, and managers are reluctant to change the content of their own organisational roles in accordance with the responsibilities and tasks of the Q-circles.

(2) The organisational role of middle managers is ambiguos. When they enrole their subordinates in the Q-circle for which they are responsible, middle managers often find themselves in conflict with the production goals for which they are responsible, because production activities are deprived of working hours. Furthermore, role conflicts may occur because those responsible for the Q-circle activities are allocated to staff functions that tresspass on the responsibilities of other subunits.

(3) A lack of operator motivation and satisfaction. The activities of the Q-circle may gradually become symbolic and meaningless, or at least percieved as such by the Q-circle participants, if the support and receptiveness to the Q-circle proposals are absent at the level of management. Thus, intrinsic motivation deteriorates and become absent. Furthermore, in some instances the Q-circle participants may be discouraged by the absence of extrinsic motivation if the outcome of the Q-circle activities are not related to the remuneration of the workforce.

In a Danish context, Wolffberg & Saabye (1988) indicates that these problems may occur frequently.[162] Furthermore, while job security and wage levels seem to stimulate the motivation of employees, the firms studied by Wolffberg & Saabye (1988) did not pay much attention to the need for training and education of the workforce in the fields of teamwork and problem solving. In addition, measures to secure feedback from the Q-circle activities to top management were hardly instituted, and the formation of strategic goals on quality was a top-down phenomenon. Finally, it appeared that the knowledge of blue collar workers as regards the present and future content of their jobs was quite limited. Compared to the Japanese way of organising Q-circle activities, the Danish principles seem to be more in line with the Anglo-Saxian tradition of conformance to requirement than with the Japanese consultative tradition of peoplism. Following Skyum & Dahlgaard (1988), the main differences between the Danish and Japanese practices of

---

162. Wolffberg & Saabye (1988) report on a study of ten Danish manufacturing firms undertaken in 1986-87. Of these ten firms, five had at least 200 employees, while five had more than 500 employees. Thus, the study is biased towards firms which in a Danish context are quite large.

Q-activities may be explained in terms of training, focus of attention and employee involvement.

(1) Training for quality. According to the Deming philosophy, training is an indispensable part of the structuring for quality. Training must take place at the levels of both management and employees, and the philosophy of quality must be diffused not only through formal training and education, but also through the training of employees by managers.

Table 14. *Training in the field of quality in Denmark and Japan*

During the last three years	Type of manager	Denmark	Japan
Average number of days spent on training and education in the field of quality	Head of central office	5.0	16.1
	Construction and product development	5.1	20.5
	Marketing	2.4	14.7
	Production	5.5	26.7
	Quality control	27.5	32.5
	Foreman or Q-circle leader	9.8	31.4
Average number of days spent by managers to train a new subordinate	Construction and product development	1.4	12.5
	Marketing	0.7	8.1
	Production	1.5	14.3
	Quality control	10.1	19.3
	Foreman (production)	2.8	17.5

Source: Skyum & Dahlgaard (1988), pp.108-09

Table 14 reveals that Danish managers spend, on average, considerably less time than Japanese managers in training their employees, and, similarly, that the amount of working days spent on training of managers are considerably smaller, with the exception of managers in quality control.[163] It is tempting to conclude that the relatively small emphasis on training for quality in Danish firms reflect a relatively slow diffusion of quality awareness which in a Danish context did not become an issue for managers and industrial policy until the late 1980s (Gjerding, 1994). However, the evidence presented in table 14

---

163. Tables 14-16 draws on Skyum & Dahlgaard (1988) who report on the empirical evidence in Dahlgaard (1987). The empirical evidence covers 41 Danish and 70 Japanese firms. Nine of the Japanese firms had more than 10,000 employees, and thus the samples are not directly comparable. Even if these nine firms are excluded, the average size of the Japanese firm in the sample is approximately 2,500 employees, which is still considerably larger than the average size of the Danish firms (Skyum & Dahlgaard, 1988, p.200).

may also reflect the fact that Danish firms, on average, allocate a relatively small amount of resources for human resource planning, including training in general (cf. Jørgensen, Lind & Nielsen, 1990, 1990a). Whichever the case, one can hardly conclude that the pursuit of quality appears as an overall organising theme for organisational behaviour, and the analysis by Wolffberg & Saabye (1988) suggest that the organisational span of attention is dispersed among competing issues of which quality management is but one. In consequence, subgoals are differentiated and communication may become less efficient due to an insufficient understanding of basic quality concepts and problems.

(2) Focus of attention. The basic understanding of quality issues and management may be measured approximately by the application of statistical methods, because these methods are vital to the understanding and improvement of quality. Table 15 applies such an approximative approach in terms of two important aspects of quality control: Building-in quality during the stage of product development and design, and applying various methods in order to facilitate the delegation of responsibility and discretionary power. As might be expected, the evidence presented in table 15 reveals some important differences between the Danish and Japanese firms in the Dahlgaard (1987) sample.

Table 15. *Quality control techniques in Denmark and Japan*

Issue	Type of technique	Denmark	Japan
Firms that employ methods in order to prevent defects	Measurement of process capability Reliability engineering	61% 56%	83% 82%
Average ranking of the importance of quality control techniques	Cause and effect chains Pareto chart Stratification Checksheet Histogram Control chart Regression analysis Scatter diagram Cause variation Sampling procedures	6 5 8 2 4 3 10 9 7 1	1 2 3 4 5 6 7 8 9 10

Source: Skyum & Dahlgaard (1988), pp.115, 210

The average Japanese firm employs to a larger degree statistical methods of quality control during the initial stages of innovation and may, in consequence, find themselves in a relatively better position to detect a number of causes for quality defects even before production has started. Furthermore, they emphasise to a larger degree relatively simple quality control techniques. This may be explained by the fact that the outcome of Q-circle activities is sensitive to the ability of workers to understand and manage quality control techniques and apply them on local discretionary behaviour. The techniques which recieve the highest ranking in the Japanese case are relatively simple and aim at removing the *causes* of defects and not the defects themselves (cf. Skyum & Dahlgaard, 1988, pp.207-11). Contrary, the techniques which recieve the highest ranking in the Danish case are relatively complex. In sum, tables 14-15 yield the impression that the management of quality is more an issue for managers than employees in the Danish case, and this may account for the problems of instituting Q-circle activities described previously.[164]

(3) Employee involvement. The previous discussion revealed that the Danish firms provide relatively less training for quality and institute the management of quality as a primarily managerial task. This implies that the degree of employee involvement in quality circles is relatively smaller in a Danish setting. Furthermore, it might be expected that the Danish Q-circle activities to a smaller degree involve employees from a wide range of productive activities within the firm.

Table 16. *Diffusion of quality circles and employee involvement*

Firm and department %-share	Denmark	Japan
Firms with Q-circles	12	97
Employees in Q-circles	1	69
Employees in Q-circles in		
- construction and product development	0.24	59.3
- marketing	0.00	54.1
- production	3.20	86.4
- quality department	7.80	72.7
- administration	0.00	67.4

Source: Skyum & Dahlgaard (1988), pp.200-01

Table 16 provides some evidence on this proposition. Although it is impossible to make any firm conclusions in the Danish case because the numbers are too small, the evidence

---

164. Furthermore, it may explain why the number of proposals generated by Danish Q-circle activities seems to be only 1/4 of the number of proposals generated in a Japanese setting. Skyum & Dahlgaard (1988, p.205) report on, respectively, 16.4 and 68.4 annual proposals. However, this observation may be explained by the differences between the two samples regarding firm size.

presented seems to indicate that employee involvement is quite small and restricted to the production and quality departments. Contrary, the Japanese evidence seems to indicate that Q-circles are used extensively across the range of task environments within the firm.[165]

The differences between the Danish and the Japanese approaches to Q-circle activities may reflect that the firms within the two samples find themselves at different points of the evolution of the management of quality. The Danish approach is based more on a top-down decision making process than on local discretionary power at multiple vertical levels where quality management is seen as a company-wide process of learning, and the Japanese experience with Q-circles had, actually, a similar starting point with the formation of task forces through top-down decision making and the subsequent emergence of Q-circles. Skyum & Dahlgaard (1988) suggest that the formation of task forces seems to be more suited for Danish firms than the formation of Q-circles since the "implementation of such groups may be the necessary precondition for a subsequent successful implementation of quality circles" (ibid., p.227).[166]

It is tempting to hypothesise a positive relationship between the implementation problems associated with advanced manufacturing technology described previously and a low speed of application of strategic quality management. The Deming philosophy provides a powerful guide to the establishment of lateral processes that may overcome the extensive burden on the incentive system, the insufficient degree of routinisation, the occurrence of information overload and audience learning, which were hypothesised conclusively in the previous section. The process of instituting Q-circle activities seem rather promising in this respect; however, a number of obstacles may prevent the potentials of quality management to be realised. The comparison between the Danish and the Japanese experiences indicates that the success of quality management in providing solutions to the flexibility-stability dilemma is sensitive to the organisational capability of the firm to provide a singleness of purpose that committs the constituent parts of the firm to quality. First and foremost, a singleness of purpose depends on the creation of mutual theories of action, i.e. the open-minded communication of goals and aspirations which creates a coherence of the different types of rationalities at work within the organisation. While these relationships are, certainly, both apparent and important within the Deming philosophy, the Danish manufacturing firms have still a long way to go before the Deming-type of organisational thinking appears at the core of the management focus of attention.

---

165. This is, of course, in accordance with the CWQC approach.
166. My translation from Danish.

234

Chapter 10
Concluding remarks

10.1. The innovation design dilemma revisited

The present study has been based on the assumption that the technology match hypothesis is valid in the sense that firms exhibit a soft variant of the technological imperative. While the technological imperative in the original formulation stated that the complexity of the production techniques is matched by the complexity of the organisational configuration, the soft variant states that this might only be partially true: A unidirectional link between technology and the organisational structure is difficult to establish across a large diversity of organisational performance, but a positive association is likely to exist in high-performance cases. This implies that high performance, measured by some surplus-cost ratio as e.g. profits or productivity, is sensitive to the establishment of a correspondance between technology and organisation.

This line of reasoning seems both intuitively appealing and absolutely trivial. How can any organisation persist, if it is not able to manage the very tools of its fabric? However, it is not obvious what we should or might mean by this question. The phrase *manage* may refer to a wide range of phenomena, ranging from the strategic decisions taken by top management to the labouring efforts undertaken by the operators at the factory floor. The present study has focussed on some processes of organisational change by which a collective of human beings arrive at a set of goals that can be pursued within the constraints of some negotiated social relations. By negotiation, the present study does not refer to deliberate negotiations between collaborating parties, but to the synergetic development of relations among people who, endowed with different skills and means of power, become dependent upon each other for the achievement of their goals. As indicated by the previous nine chapters, it is far from trivial to obtain a satisfactory explanation of these phenomena.

The issue of establishing a correspondance between technology and organisation is particularly interesting in cases of change in one or both factors. Put in general terms, change may be initiated from within or without the organisation, but is in any case driven from within by the organisation members. In consequence, change does not only imply a change in the way that things are done, but also in the way that things are percieved, since it is impossible to concieve a change of human activities without a change in the perceptions held by these humans. These perceptions may drift in different directions, and thus it is paramount that the implementation of a new state of affairs is based upon the formation of a new set of negotiated relationships which can coordinate the activities of the collective. From a structural-functionalist point of view, the present study has defined the set of negotiated relationships in terms of organisational routines that constitute a social equilibrium. The social equilibrium is dynamic in the sense that it responds to performance gaps which occur because the intra- and extra-organisational environments are changing. Thus, organisational structures are faced with a flexibility-stability dilemma: While they must remain stable in order to achieve a certain level of performance, they must at the same time be flexible in order to adapt to changing circumstances. These changes do not come about in an automatic fashion as manna from heaven since changes in the extraorganisational environment are initiated and instituted by the very organisations that constitute the environment, and in many cases organisations even design themselves in order to change.

From this point of view, the present study has delved into *three research questions*: How can the flexibility-stability dilemma be described in terms of the behavioural theory of the firm and the contingency theory of organisational action? How can this description be interpreted from the point of view of the economics of innovation? How may this interpretation contribute to the explanation of the contemporary organisational changes most frequently associated with the diffusion and implementation of microelectronics and information technology?

10.2. Condensed answers to broad questions

Each of these questions are quite broad, and so is the argument presented in chapters 2-9. The questions have been answered by outlining a number of perspectives on the flexibility-stability dilemma derived from (1) behavioural and contingency theorising, (2) neo-behavioural and evolutionary theorising, and (3) innovation economics. While section 1.5 contains a more comprehensive summary of the analysis undertaken in chapters 2-9, the present and final section constrains itself to presenting some condensed answers which in some instances may give birth to more questions than they finalise.

How can the flexibility-stability dilemma be described in terms of the behavioural theory of the firm and the contingency theory of organisational action?

The behavioural interpretation of the flexibility-stability dilemma presented in this volume rests, of course, on the concepts of bounded rationality and satisficing decision making which are utilised in the explanation of how organisational behaviour depends on the determination of the focus of attention. The focus of attention changes as the outcome of organisational learning which may be more or less programmed and is stimulated by organisational slack that tends to trade off the influence of local rationality. Organisational learning takes place as problemistic search associated with reproductive or productive problem-solving. Innovation is, primarily, stimulated by the number of attention centres and the differentiation of subgoals, and impeded by the persistence of subgoals. Innovation is, furthermore, negatively associated with the degree of satisfaction: The greater the level of satisfaction, the lower the rate of change, and *vice versa*. Finally, innovation is to a higher degree initiated by those parts of the organisation which deal with the extraorganisational environment, and to a lower degree by those parts of the organisation which are engaged in the most detailed performance programs.

The last observation is utilised in the contingency theorising on organisational action, in two respects. *First*, organisations differentiate their organisational design across subunits in order to deal with the various parts of the extraorganisational environment. This implies that the organisation is faced with a problem of integration of its subsequent parts. *Second*, organisations tend to apply a combination of closed- and open-systems rationality in their design in order to manage contingencies creating uncertainty. This implies that the organisation is faced with a conflict between computational and non-computational rationality, reflected in a diversity of intraorganisational perspectives.

Both problems entail the danger of organisational conflict, the resolution of which becomes effective only if the distance between the organisation members' perceptions is small and integration performed at the appropriate levels of influence. The difficulty of coordination is positively related to the importance of contingency, and thus to the frequency and amount of communication entailed in decision making. Consequently the organisation is faced with uncertainty to the extent that there exists a gap between the knowledge of the organisation members and the knowledge necessary to cope with contingencies.

How can this description be interpreted from the point of view of the economics of innovation?

The basic assumption of the interpretation from an innovation economics point of view is that solutions to the flexibility-stability dilemma involve both technical and organisational innovation. Within innovation economics, learning associated with innovation is most often described in terms of technical change, i.e. product and process innovation, but it is increasingly recognised that technical change is linked to organisational innovation, not only at aggregate levels as in the emerging national system of innovations tradition, but also at the level of the organisation. The interpretation of the behavioural and contingency approach to the flexibility-stability dilemma is undertaken in two steps: *First*, the description of organisational learning is qualified in terms of neo-behavioural incidents where a learning cycle breaks down, and reinterpreted in terms of learning concepts found within an institutional approach to innovation economics. *Second,* the flexibility-stability dilemma is discussed in terms of patterns of innovative behaviour derived from a broad range of theoretical and empirical work within the field of innovation economics. These two steps imply a broad definition of the field of innovation economics, since they encompass evolutionary and institutional theorising alongside the economics of innovation.

Changes of the focus of attention, and thus changes of the constraints which defines the boundaries of organisational rationality and the degree of satisfaction, depends on the type of learning involved. The neo-behavioural incidents occur as learning which is role-constrained, audience-like, superstitious or ambigous. These types of learning pathologies reflect that the outcome of previous intra- and extraorganisational action must be subjected to interpretation within the intraorganisational set of negotiated relationships. Furthermore, they may to some extent be present in the cases of learning by producing, searching and exploring, where learning by producing refers to learning by doing in the Arrow (1962) sense, learning by using in the Rosenberg (1982) sense, and learning by interacting in the Lundvall (1985) sense. These interfaces may be described in terms of single-loop, double-loop and deutero learning.

From an evolutionary point of view, the set of negotiated relationships may be described in terms of a set of organisational routines that defines the path along which organisational learning takes place. The organisational routines provide the organisation members with a socially ordered context in which individual learning takes place and translates into organisational learning. This implies that learning is an interactive process subjected to behavioural regularities. If one imagines a continuum of interaction, one may find learning by searching and exploring at the high-frequency end, while learning by producing is located within the middle range. However, this should not be interpreted as if the difficulties of initiating and sustaining innovation from a flexibility-stability dilemma point of view are positively correlated with the frequency of interaction, since the number of difficulties depend on the type of learning pathologies involved.

Within innovation economics, the process of technical innovation has mostly been described as a linear progression of sequential stages. However, the flexibility-stability approach argues that innovation, whether technical or organisational, takes place through iterative feedbacks between the stages of initiation and implementation. In consequence, a reinterpretation within innovation economics must emphasise the process of feed-back. The chain-linked model (Kline & Rosenberg, 1986) offers an alternative to the linear models of technical innovation, which emphasises the processes of feedback that takes place along a central core of progressive stages denoting the evolution of a dominant design. However, this approach to the analysis of technical innovation abstracts from the type of organisational conflicts that may accompany technical innovation. In order to encompass organisational conflicts, the resolution of organisational conflicts, and thus the types of organisational changes which accomodate technical innovation, sociological complex real-type models of the accumulation and envelope type (Christensen, J.F., 1992) may be used in order to elaborate on the reasoning of the chain-linked model.

Another approach to the interpretation of the flexibility-stability approach may be to focus on the types of strategies for technical innovation that firms employ. Freeman (1982) describes six types of strategies, which are defined in terms of the competencies that form the core of the firm strategy. For instance, while both an opportunistic and imitative strategy rely on scientific and technical information, an opportunistic strategy emphasises long-range forecasting and production planning while an imitative strategy emphasises production engineering and quality control. The Freeman approach may be augmented by focussing on the match between the intra- and extraorganisational task environments in terms of the forces which generate and integrate the technical capabilities of the firm, and it may be argued that "successful firms operate within some sort of harmonious equilibrium of these forces" (Burgelman & Rosenbloom, 1989, p.20). To the extent that such an equilibrium cannot be created and sustained, the flexibility-stability dilemma cannot be resolved at the level of satisfaction. There is still much work to be done in order to describe the Freeman typology from the perspective of the Burgelman & Rosenbloom cultural evolutionary theorising. Furthermore, important insights may also be obtained by combining the spectrum of innovation processes suggested by Cooper (1983) with the sectoral patterns of innovation proposed by Pavitt (1984).

The main conclusion to be drawn from these considerations is that the analysis of technical progress must rely on an understanding of technical capabilities as something that is generated by the co-evolution of firms and their environments. That is, the generation and integration of technical capabilities is determined by the interaction between the intra- and extraorganisational task environments. This interaction may be described in terms of a path for technical progress, which constitutes a paradigm in the Kuhnian sense. The notion of techno-economic paradigms suggested by Freeman & Perez (1988) implies that the path of technical progress is cumulative in the sense that it defines a number of constraints and opportunities for future search. The techno-economic paradigm reflects an accumulation of endowed resources and knowledge which exerts an exclusion effect on solutions and implies complementaries between the economic agents at the levels of the firm, the sector, the production system, and the economy. Consequently, only those firms which are able to adhere to the prevailing techno-economic paradigm will survive in the long run.

How may this interpretation contribute to the explanation of the contemporary organisational changes most frequently associated with the diffusion and implementation of microelectronics and information technology?

The answer to the last question takes, as its point of departure, the techno-economic point of view and presupposes that the performance of contemporary manufacturing firms is closely associated with the ability of the firm to exploit the opportunities offered by advanced manufacturing technology. However, this should not be interpreted as if the most successful firms are those able to make the most capital intensive investments. Instead, it should be recognised that long-term survival is determined by the ability of the firm to obtain a *match* between its technical and organisational capabilities.

This point of view, which reflects the soft variant of the technological imperative, stresses the importance of human resource management, the simplification of organisational hierarchies, and the achievement of an agreement between management and labour on technical change in order to balance technical and organisational innovation. At the present stage of the emerging new techno-economic paradigm, the balance between technical and organisational innovation is especially important because the successful implementation of advanced manufacturing technology seems to imply a trend towards organic principles of organisation. The trend towards organic principles of organisation poses a huge challenge to the modern firm: Organisational changes are more difficult to initiate and institute than technical change since they are more like to stimulate organisational conflict; furthermore, organisations learn faster about technical issues than organisational issues because knowledge about organisational issues is, to a larger degree, tacit and, to a smaller degree, facilitated by organisational arrangements. In consequence, the modern firm encounters a problem of maintaing *organisational coherence*. This problem may become enhanced because the implementation of advanced manufacturing technology implies an increase in the skills content of the workforce which creates a larger degree of organisational complexity in terms of the number of occupational specialities and the degree of professionalism in each. As discussed previously in chapter 2, a high degree of complexity enhances the organisation's receptiveness to innovation. A larger degree of receptiveness to innovation may increase the amount of opportunistic surveillance relative to problemistic search. Proactive perceptions of performance gaps may be stimulated, and the intraorganisational theories of action may, to a larger degree, become embedded with relatively more solutions looking for a problem than problems looking for a solution. A situation like this is likely to generate creative tensions related to an increased diversity of feedbacks and lateral processes by which the mutual theories of action are created and maintained. In order to solve these problems, the modern firm may promote stress or slack innovation relative to innovation occasioned by neccessity or programmed action.

Quality management provides the modern firm with an opportunity to solve the flexibility-stability dilemma by creating mutual theories of action. In the case of the *Deming philosophy*, we are dealing with a kind of quality management that goes beyond the issue of quality *control* mentioned previously in relation to the imitative strategy described by Freeman (1982). The Deming approach implies that the firm apply long-term global rationality within a cooperative framework where the management provides a uniform direction of purpose, and where the responsibilities of the employees are aligned with local discretion. Furthermore, these organisational principles ought to be supplemented by the training necessary within a corporate atmosphere of psychological job security. Thus, the Deming approach are in accordance with the principles of visibility previously described in chapter 5. However, the Wollfberg & Saabye (1988) survey described in section 9.2 reveals a number of real-life problems throughout the industrialised world contradictory to the kind of visibility advocated in chapter 5. This points to the fact that the generation of mutual theories of action through quality management still has a long way to go before it becomes effective. Presumably, some new initiatives in the field of management practices are needed, such as the prescriptions implied by the European quality award or the analytical approach of Business Process Re-engineering (BPR).

Although the three research questions have been answered to a larger extent that indicated by the condensed answers given above, it may be argued that the present volume only to a limited extent has succeeded in bridging the gap between organisation theory and innovation economics.

First, the organisation theories applied are classic. The classic approach reflects my interest for understanding the roots of the after-war organisation theory within a general evolutionary and a specific dynamic structural-functionalist framework. This approach seems valid because the present volume applies organisation theory to a field of economic thinking which is structural-functionalist and increasingly oriented towards evolutionary thinking. However, a number of new strands within organisation theory may provide the bridging attempt with a more dynamic approach, e.g. theories on corporate culture, business process re-engineering and social constructionism. Thus, we are faced with an important future line of research which becomes feasible in those field of innovation economics which deal with institutional and sociological matters, such as the institutional approach described in chapter 6 and the complex real-type models described in chapter 7. This line of thinking is increasingly becoming important in the field of innovation economics, e.g. under the heading of *the learning economy* (Foray & Lundvall, 1996).

241

Second, a number of research questions are still left open. The intersections of neo-behavioural and evolutionary concepts of learning merit a more subtle analysis than undertaken in section 6.5. The analysis of the relationship between the Cooper (1983) clusters across the Pavitt (1984) categories described in section 7.2 is yet another very interesting issue left for future research. The application in chapter 9 of ideal-type Japanese principles of management may be questionable since we are dealing with normative prescriptions seldom observed in real-life firms; the same argument applies to the Lucas definition of advanced manufacturing technology invoked by chapter 8. Finally, the opportunities provided by quality management calls for a research project of its own, and the present volume fails to cover the literature proliferating in this field.

References

Adler, P.S. (1989), "Technology Strategy: A Guide to the Literatures", *Research on Technological Innovation, Management and Policy*, Vol.4, pp.25-151.

Adler, P.S. (1990), "Managing High Tech Processes: The Challenge of CAD/CAM", pp.188-215 in Glinow, M.A. von & Mohrmann, S.A. (eds.), *Managing Complexity in High Technology Organizations*, New York: Oxford University Press.

Alchian, A.A. & Demsetz, H. (1972), "Production, Information Costs, and Economic Organization", *American Economic Review*, Vol.62, pp.777-95.

Amabile, T.M. (1988), "From Individual Creativity to Organizational Innovation", pp.139-66 in Grønlaug, K. & Kaufmann, E. (eds.), *Innovation: A Cross-Disciplinary Perspective*, Oslo: The Norwegian University Press.

Aoki, M. (1984), "Aspects of the Japanese Firm", pp.3-43 in Aoki, M. (ed.), *The Economic analysis of the Japanese Firm*, Amsterdam: North-Holland.

Aoki, M. (1990), "A New Paradigm of Work Organization and Co-ordination?", pp.267-93 in Marglin, S.A. & Schor, J.B. (eds.), *The Golden Age of Capitalism*, Oxford: Clarendon Press.

Aoki, M. (1991), "Global Competition, Firm Organization and Total Factor Productivity. A Comparative Micro Perspective", pp.419-25 in OECD, *Technology and Productivity. The Challenge for Economic Policy*, Paris: OECD.

Argyris, C. & Schön, D.A. (1978), *Organizational Learning: A Theory of Action Perspective*, Reading, Mass.: Addison-Wesley.

Arrow, K.J. (1962), "The Economic Implications of Learning by Doing", *Review of Economic Studies*, Vol.29, pp.155-73.

Arthur, W.B. (1988), "Competing technologies: an overview", pp.590-607 in Dosi et al. (1988).

Bain, J.S. (1949), "A Note on Pricing in Monopoly and Oligopoly", *American Economic Review*, Vol.39, pp.448-64.

Bain, J.S. (1956), *Barriers to New Competition*, Cambridge, Mass.: Harvard University Press.

Banke, P. & Clematide, B. (1989), *Ny teknik - nye job* (New technology - new jobs), København: Nordisk Ministerråd.

Bateson, G. (1942), "Social Planning and the Concept of Deutero-Learning", pp.159-176 in Bateson, G. (1972), *Steps to an Ecology of Mind*, New York: Ballantine Books.

Baumol, W.J. (1959), *Business Behavior, Value and Growth*, New York: MacMillan, 2nd Edition, New York: Harcourt, Brace & World, 1967.

Bessant, J. (1989), "Flexible Manufacturing: Yesterday, Today, Tomorrow", in Bolk et al. (eds.), *Implementing flexible manufacturing*, Amsterdam: Eburon Publications.

Bessant, J. & Buckingham, J. (1989), "Implementing integrated technology", *Technovation*, Vol.9, pp.321-36.

Blaug, M. (1978), *Economic Theory in Retrospect*, 3rd Edition, New York: Cambridge University Press.

Brown, R.H. (1978), "Bureaucracy as Praxis: Toward a Political Phenomenology of Formal Organizations", *Administrative Science Quarterly*, Vol.23, pp.365-82.

Bruzelius, L.H. & Skärvad, P.-H. (1979), *Integrerad företagsadministration* (Integrated Business Administration), Lund: Studentlitteratur.

Burgelman, R.A. & Sayles, L.R. (1986), *Inside Corporate Innovation*, New York: The Free Press & Collier Macmillan.

Burgelman, R.A. & Rosenbloom, R.S. (1989), "Technology Strategy: An Evolutionary Process Perspective", *Research on Technological Innovation, Management and Policy*, Vol.4, pp.1-23.

Burns, T. & Stalker, G.M. (1961), *The Management of Innovation*, London: Tavistock.

Butler, R. (1991), *Designing Organizations*, Chatham: Routledge.

Chamberlin, E. (1933), *The Theory of Monopolistic Competition*, 8th Edition, Cambridge, Mass.: Harvard University Press, 1962.

Cavestro, W. (1986), "Automation, work organization and skills: the case of numerical control", *Automatica*, Vol.22, reprinted in Cahiers IREP-D, *Industrial flexibility and work*, Grenoble: The Institute of Economic Research and Development Planning.

Chandler, A.D. (1992), "Organizational Capabilities and the Economic History of the Industrial Enterprise", *Journal of Economic Perspectives*, Vol.6, pp.79-100.

Chiaromonte, F. & Dosi, G. (1993), "The micro foundations of competitiveness and their macroeconomic implications", pp.107-34 in Foray, D. & Freeman, C. (eds.), *Technology and the Wealth of Nations*, London: OECD & Pinter Publishers.

Christensen, J. (1992), "Ny teknologi - køb, implementering og anvendelse" (New technology - purchase, implementation and use), pp.219-29 in Parum, E. & Jensen, F.L. (eds.), *Inspiration til aktivt bestyrelsesarbejde* (Inspiration to proactive management at the board - experiences from the Danish industry), Greve: Børsen Bøger.

Christensen, J.F. (1992), *Produktionnovation - proces og strategi* (Product Innovation - Process and Strategy), Hvidovre: Handelshøjskolens Forlag.

Christensen, J.L. (1992), "The Role of Finance in National Systems of Innovation", pp.146-68 in Lundvall (1992).

Christensen, P.R., Andersson, J. & Blenker, P. (1992), *Industriens brug af underleverandører* (Industrial producer-supplier cooperation), København: Industri- og Handelsstyrelsen.

Clark, P. & Starkey, K. (1988), *Organization Transitions and Innovation-Design*, London: Pinter.

Clark, P. & Staunton, N. (1989), *Innovation in Technology and Organization*, London: Routledge.

Clegg, S.R. (1990), *Modern Organizations*, Trowbridge: Sage.

Coase, R.H. (1937), "The Nature of the Firm", *Econometrica*, Vol.4, pp.386-405.

Cohen, M.D., March, J.G. & Olsen, J.P. (1976), "People, Problems, Solutions and the Ambiguity of Relevance", pp.24-37 in March, J.G. & Olsen, J.P. (eds.), *Ambiguity and Choice in Organizations*, Oslo: Universitetsforlaget.

Coombs, R., Saviotti, P. & Walsh, V. (1987), *Economics and Technical Change*, Hong Kong: MacMillan.

Cooper, R.G. (1983), "The new product process: an empirically-based classification scheme", *R&D Management*, Vol.13, pp.1-13.

Cross, J.G. (1983), *A theory of adaptive economic behavior*, Cambridge, Mass.: Cambridge University Press.

Cyert, R.M. & Hedrick, C.L. (1972), "Theory of the Firm: Past, Present, and Future: An Interpretation", *The Journal of Economic Literature*, Vol.10, pp.398-412.

Cyert, R.M. & March, J.G. (1963), *A Behavioral Theory of the Firm*, Englewood Cliffs: Prentice-Hall.

Dahlgaard, J.J. (1987), *En sammenlignende undersøgelse af kvalitetsstyringens metoder og principper i Japan, Sydkorea og Danmark* (A comparative analysis of the methods and principles of quality control in Japan, South Korea and Denmark), Århus: Handelshøjskolen.

Deming, W.E. (1982), *Quality, Productivity, and Competitive Position*, Cambridge, Mass.: Centre for Advanced Engineering Study, MIT.

Deming, W.E. (1986), *Out of the Crisis*, Cambridge, Mass.: The MIT Press.

Dertouzos, M.L., Lester, R.K. & Solow, R.M. (1989), *Made in America*, Cambridge, Mass.: The MIT Press.

Donaldson, L. (1985), *In Defence of Organization Theory*, Cambridge: Cambridge University Press.

Dosi, G. (1982), "Technological paradigms and technological trajectories", *Research Policy*, Vol.11, pp.147-62.

Dosi, G. (1988), "The nature of the innovation process", pp.221-38) in Dosi et al. (1988).

Dosi, G. et al. (1988), *Technical Change and Economic Theory*, London: Pinter Publishers.

DS (1988), *Kvalitet - Ordliste, Quality - Vocabulary, DS/ISO 8402*, København: Dansk Standardiseringsråd.

Dunning J.H. (1990), *The Governance of Japanese and U.S. Manufacturing Affiliates in the U.K.: Some Country Specific Differences*, paper presented at the Workshop on the Organization of Work and Technology: Implications for International Competitiveness, European Institute of Advanced Management Studies (EIASM), Brussels, May 31-June 1.

Edge, J. (1990), "Quality Improvement: Lessons for Management", in Lock, D. & Smith, D.J. (eds.), *Gower Handbook of Quality Management*, Worcester: Gower.

Edge, J. (1990a), "Quality Improvement Activities and Techniques", in Lock, D. & Smith, D.J. (eds.), *Gower Handbook of Quality Management*, Worcester: Gower.

Edquist, C. & Jacobsson, S. (1988), *The Global Diffusion of New Technology in the Engineering Industry*, Oxford: Basil Blackwell.

Edquist, C. & Lundvall, B.-Å. (1993), "Comparing the Danish and Swedish Systems of Innovation", pp.265-98 in Nelson, R.R. (ed.), *National Innovation Systems*, New York: Oxford University Press.

Feldman, M.S. & March, J.G. (1981), "Information in Organizations as Signal and Symbol", *Administrative Science Quarterly*, Vol.26, pp.171-86.

Ferdows, K. & Meyer, A. De (1987), *Manufacturing Futures Survey*, INSEAD Working Papers, Fontainebleau.

Foray, D. & Lundvall, B.-Å. (1996), *The knowledge-based economy: From the economics of knowledge to the learning economy*, paper presented at the conference on Creativity, Innovation and Job Creation, organised by the OECD and the Norwegian Ministry of Education, Research and Church Affairs, Oslo, January 11-12.

Freeman, C. (1982), *The Economics of Industrial Innovation*, 2nd ed., London: Frances Pinter.

Freeman, C. (1987), *Technology and Economic Performance: Lessons from Japan*, London: Pinter Publishers.

Freeman, C. (1988), "Japan: a new national system of innovation", pp.330-48 in Dosi et al. (eds.), *Technical Change and Economic Theory*, London: Pinter Publishers.

Freeman, C. (1992), *The Economics of Hope*, London: Pinter Publishers.

Freeman, C. (1992a), "Innovation, changes of techno-economic paradigm and biological analogies in economics", pp.121-42 in Freeman (1992). Earlier versions published in *T T T Katsaus*, No.4, 1989 and *Revue Économique*, Vol.42, pp.211-32.

Freeman, C. (1992b), "The human use of human beings and technical change", paper presented at the European Foundation for the Improvement of Living and Working Conditions Conference in The Hague, October 9, 1990, reprinted at pp.174-89 in Freeman (1992).

Freeman, C. & Perez, C. (1988), "Structural crisis of adjustment, business cycles and investment behaviour", pp.38-66 in Dosi, G. et al. (eds.), *Technical Change and Economic Theory*, London: Pinter Publishers.

Galbraith, J.R. (1973), *Designing Complex Organizations*, Reading, Mass.: Addison-Wesley.

Galbraith, J.R. (1977), *Organization Design*, Reading, Mass.: Addison-Wesley.

Garratt, B. (1983), "The power of Action Learning", pp.23-53 in Pedler, M. (ed.), *Action Learning in Practice*, Aldershot: Gower.

Gay, P. (1983), "Action Learning and organisational change", pp.153-64 in Pedler, M. (ed.), *Action Learning in Practice*, Aldershot: Gower.

Gitlow, H.S. & Gitlow, S.J. (1987), *The Deming Guide to Quality and Competitive Position*, Englewood Cliffs: Prentice-Hall.

Gergen, K.J. (1992), "Organization Theory in the Postmodern Era", pp.207-26) in Reed, M. & Hughes, M. (eds.), *Rethinking Organization*, London: Sage.

Gjerding, A.N. (1990), "Danske virksomheders anvendelse af ny teknologi" (The use of new technology in Danish manufacturing), pp.373-402 in Gjerding et al. (1990a).

Gjerding, A.N. (1991), "Produktivitetsparadokset og den nye teknologi" (The productivity paradox and the new technology), *Nationaløkonomisk Tidsskrift*, Vol.129, pp.54-66.

Gjerding, A.N. (1991a), "Produktivitet og konkurrenceevne i SMEC" (The modelling of productivity and international competitiveness in SMEC), *Samfundsøkonomen*, No.5, pp.25-30.

Gjerding, A.N. (1992), "Work Organisation and the Innovation Design Dilemma", pp.95-115 in Lundvall (1992).

Gjerding, A.N. (1992a), "Midt i et organisatorisk vadested", (Midstream Management), *Ledelse i dag*, Spring, No.6, pp.16-22.

Gjerding, A.N. (1994), "Denmark", pp.71-108 in Barker, B. (ed.), *Quality Promotion in Europe*, Cambridge: Gower Publishing.

Gjerding, A.N. (1994a), "Quality Mangement of Public Transportation by Bus", pp.477-88 in Lahrmann, H. & Pedersen, L.H. (eds.), *Trafikdage på AUC* (Traffic Days at Aalborg University), Aalborg: Transportrådet & Trafikforskningsgruppen, Aalborg Universitet.

Gjerding, A.N. & Kallehauge, L. (1990), "Produktivitetsudviklingen i ADAM og SMEC" (The modelling of productivity in ADAM and SMEC), pp.53-90 in Gjerding et al. (1990a).

Gjerding, A.N. & Lauridsen, B. (1995), "Konflikten mellem menneske og strategi" (The conflict between humans and strategies), *Samfundsøkonomen*, No.1, pp.31-37.

Gjerding, A.N. & Lundvall, B.Å. (1990), "Teknisk fornyelse og produktivitetsudvikling i danske industrivirksomheder 1984-89" (Technical innovation and the growth of productivity in the Danish industry 1984-89), pp.403-514 in Gjerding et al. (1990a).

Gjerding, A.N. & Lundvall, B.Å. (1992), "Teknisk fornyelse og konkurrenceevne - udfordringer for bestyrelsen" (Technical change and competitiveness - challenges to the board), pp.165-80 in Parum, E. & Jensen, F.S. (eds.), *Inspiration til aktivt bestyrelsesarbejde - konkrete erfaringer fra dansk erhvervsliv* (Inspiration to proactive management at the board - experiences from the Danish industry), Greve: Børsen Bøger.

Gjerding, A.N. & Madsen, P.T. (1992), "I brudfladen mellem menneske og teknik..." (The human-machine interface), pp.71-84 in Kohl, B. & Melander, P. (eds.), *Ledelse og informationstekno-logi - 90'ernes udfordring* (Management and information technology - the challenge of the 1990s), Charlottenlund: Jurist- og Økonomforbundets Forlag.

Gjerding, A.N., Johnson, B., Kallehauge, L., Lundvall, B.-Å. & Madsen, P.T. (1988), *Produk-*

tivitet og international konkurrenceevne (Productivity and International Competitiveness), Århus: Aalborg Universitetsforlag.

Gjerding, A.N., Johnson, B., Kallehauge, L., Lundvall, B.-Å. & Madsen, P.T. (1990), *Den forsvundne produktivitet*, Charlottenlund: Jurist- og Økonomforbundets Forlag. Published in English as *The Productivity Mystery*, Charlottenlund: DJØF Publishing, 1992.

Gjerding, A.N., Johnson, B., Kallehauge, L., Lundvall, B.-Å. & Madsen, P.T. (1990a), *Jagten på den forsvundne produktivitet. Bilagsrapport* (The Quest for the Lost Productivity. Supplementary Report), Charlottenlund: Jurist- og Økonomforbundets Forlag.

Gupta, Y.P. (1988), "Organizational issues of flexible manufacturing systems", *Technovation*, Vol.8, pp.255-69.

Hacche, G. (1979), *The Theory of Economic Growth*, Trowbridge & Esher: MacMillan.

Hage, J. & Aiken, M. (1969), "Routine Technology, Social Structure, and Organization Goals", *Administrative Science Quarterly*, Vol.14, pp.366-76.

Hage, J. & Aiken, M. (1970), *Social Change in Complex Organizations*, Clinton, Mass.: Random House.

Hagedoorn, J. (1990), "Organizational modes of inter-firm cooperation and technology transfer", *Technovation*, Vol.10, pp.17-28.

Hall, R. & Hitch, C. (1939), "Price Theory and Business Behavior", *Oxford Economic Papers*, Vol.2, pp.12-45.

Haywood, B. & Bessant, J. (1987), *The Swedish Approach to the Use of Flexible Manufacturing Systems*, Occasional Paper No.3, Brighton: Innovation Research Group, Brighton Polytechnic.

Hedberg, B. (1981), "How organizations learn and unlearn", pp.3-27 in Nystrom, P.C. & Starbuck, W.H. (eds.), *Handbook of Organizational Design*, Oxford: Oxford University Press.

Hickson, D.J., Pugh, D.S. & Pheysey, D.C. (1969), "Operations Technology and Organization Structure: An Empirical Reappraisal", *Administrative Science Quarterly*, Vol.14, pp.378-97.

Holbek, J. (1988), "The Innovation Design Dilemma: Some Notes on Its Relevance and Solutions", pp.253-77 in Grønlaug, K. & Kaufmann, E. (eds.), *Innovation: A Cross-Disciplinary Perspective*, Oslo: Norwegian University Press.

Högberg, L. (1981), *Industriel innovation* (Industrial innovation), Århus: Jydsk Teknologisk Instituts Forlag.

Industri- og Handelsstyrelsen (1988), *Spredningen af informationsteknologi i dansk industri 1987* (The diffusion of information technology in the Danish industry 1987), København: Industriministeriet (The Ministry of Industry).

Insley, S. (1989), "The Honda Way: An Innovative Approach to Management and Production", pp.123-33 in Lundstedt, S.B. & Moss, T.H. (eds.), *Managing Innovation and Change*, Dordrecht: Kluwer Academic Publishers.

Jelinek, M. & Goldhar, J.D. (1983), "The Interface between Strategy and Manufacturing Technology", *Colombia Journal of World Business*, reprinted on pp.401-16 in Tushman, M.L. & Moore, W.L. (1988), *Readings in the Management of Innovation*, 2nd Edition, Cambridge, Mass.: Ballinger.

Jensen, E. (1992), *Samarbejde eller konkurrence?* (Cooperation or competition?), Aalborg: Aalborg Universitetsforlag.

Jensen, J.O. & Christensen, O. (1993), *Økonomistyring og budgettering* (Financial control and budgetting), Gylling: Systime.

Johnson, B. (1981), *Aktuelle tendenser i den økonomiske politik under krisen. Del I. Den nyliberale tendens* (Current trends in the economic policy during the crisis. Part I. The neo-liberal trend), Aalborg: Aalborg Universitetsforlag.

Johnson, B. (1992), "Institutional Learning", pp.23-44 in Lundvall (1992).

Johnson, B. & Lundvall, B.-Å. (1992), "Closing the Institutional Gap?", *Revue d'Economie Industrielle*, No.59, 1er Trimestre.

Joiner, B.L. & Scholtes, P.R. (1988), "The manager's new job", pp.29-36 in Chase, R.L. (ed.), *Total Quality Management*, Exeter: IFS Publications & Springer Verlag.

Jørgensen, H., Lind, J. & Nielsen, P. (1990), *Personale, planlægning og politik* (Human resource development and policy), Aalborg: ATA-forlaget.

Jørgensen, H., Lind, J. & Nielsen, P. (1990a), *Personaleplanlægning og arbejdsmarkedsservice* (Human ressource planning and the institutional agents at the labour market), Aalborg: ATA-Forlaget.

Kallehauge, L.E. (1990), "Spredningen af numerisk styrede værktøjsmaskiner" (The diffusion of CNC machinery), pp.321-71) in Gjerding et al. (1990a).

Kamien, M.I. & Schwartz, N.L. (1982), *Market structure and innovation*, Cambridge: Cambridge University Press.

Kay, N. (1979), *The Innovating Firm*, Thetford: MacMillan.

Kay, N. (1988), "The R&D function: corporate strategy and structure", pp.282-94 in Dosi et al. (1988).

Khandwalla, P.N. (1974), "Mass Output Orientation of Operations Technology and Organizational Structure", *Administrative Science Quarterly*, Vol.19, pp.74-97.

Kline, S.J. & Rosenberg, N. (1986), "An Overview of Innovation", pp.275-305 in Landau, R. & Rosenberg, N. (eds.), *The Positive Sum Strategy*, Washington D.C.: National Academy Press.

Knudsen, C. (1991), *Økonomisk metodologi* (Economic Methodology), Charlottenlund: Jurist-og Økonomforbundets Forlag.

Kogut, B. (1991), "Contry Capabilities and the Permeability of Borders", *Strategic Management Journal*, Vol.12, pp.33-47.

Kristensen, A. (1992), *Industriel innovation i Danmark* (Industrial innovation in Denmark), Aalborg: Aalborg Universitetsforlag.

Lazonick, W. & Brush, T. (1985), "The 'Horndahl Effect' in Early U.S. Manufacturing", *Explorations in Economic History*, Vol.22, pp.53-96.

Lawrence, P.R. & Lorsch, J.W. (1967), *Organization and Environment. Managing Differentiation and Integration*, Georgetown, Ontario: Irwin-Dorsey, 1969.

Leibenstein, H. (1966), "Allocative Efficiency vs. 'X-efficiency'", *American Economic Review*, Vol.56, pp.392-415.

Leibenstein, H. (1987), *Inside the Firm. The Inefficiencies of Hierarchy*, Cambridge, Mass.: Harvard University Press.

Leonard-Barton, D. (1988), "Implementation as mutual adaptation of technology and organization", *Research Policy*, Vol.17, pp.251-67.

Leonard-Barton, D. (1990), "Implementing New Production Technologies: Exercises in Corporate Learning", pp.160-87 in Glinow, M.A. von & Mohrmann, S.A. (eds.), *Managing Complexity in High Technology Organizations*, New York: Oxford University Press.

Levinthal, D. (1991), "Random Walks and Organizational Mortality", *Administrative Science Quarterley*, Vol.36, pp.397-420.

Levinthal, D. (1992), "Surviving Schumpeterian Environments: An Evolutionary Perspective", *Industrial and Corporate Change*, Vol.1, pp.427-43.

Lincoln, J.R. (1990), *Work Organization in Japan and the United States*, paper presented at the Workshop on the Organization of Work and Technology: Implications for International Competitiveness, European Institute of Advanced Management Studies (EIASM), Brussels, May 31-June 1.

Littler, D.A. & Sweeting, R.C. (1984), "Business innovation in the UK", *R&D Management*, Vol.14, pp.1-9.

Lundberg, E. (1961), *Produktivitet och räntabilitet* (Productivity and Profitability), Stockholm: Studieförbundet Näringsliv och Samhälle.

Lundvall, B.-Å. (1985), *Product Innovation and User-Producer Interaction*, Aalborg: Aalborg University Press.

Lundvall, B.-Å. (1988), "Innovation as an interactive process: from user-producer interaction to the national system of innovation", pp.349-69 in Dosi et al. (1988).

Lundvall, B.-Å. (1992), ed., *National Systems of Innovation*, London: Pinter Publishers.

Lundvall, B.-Å. (1992a), "Introduction", pp.1-19 in Lundvall (1992).

Madsen, P.T., Gjerding, A.N., Johnson, B. & Kallehauge, L. (1993), "Produktivitetens mange determinanter - svar på en kritik" (The multitude of productivity determinants - reply), *Samfundsøkonomen*, No.5, pp.29-35.

March, J.G. & Olsen, J.P. (1976), "Organizational Choice under Ambiguity", pp.10-23 in March, J.G. & Olsen, J.P. (eds.), *Ambiguity and Choice in Organizations*, Oslo: Universitetsforlaget.

March, J.G. & Olsen, J.P. (1976a), "Organizational Learning and the Ambiguity of the Past", pp.54-68 in in March, J.G. & Olsen, J.P. (eds.), *Ambiguity and Choice in Organizations*, Oslo: Universitetsforlaget.

March, J.G. & Simon, H.A. (1958), *Organizations*, New York: John Wiley & Sons.

Marris, R. (1964), *The Economic Theory of Managerial Capitalism*, Glencoe: The Free Press.

Marris, R. & Mueller, D.C. (1980), "The Corporation, Competition, and the Invisible Hand", *Journal of Economic Literature*, Vol.18, pp.32-63.

Melander, P. (1994), *Økonomistyring og budgettering som ledelsesform* (Financial control and budgetting as management practice), Viborg: Handelshøjskolens Forlag.

Meyer, A. De & Ferdows, K. (1991), "Removing the Barriers in Manufacturing", pp.198-212 in EIASM, *Management and New Production Systems*, proceedings from the 3rd international production management conference on management and new production systems, Göteborg, May 27-28, organised by European Institute of Advanced Studies in Management (EIASM) and University of Göteborg, Department of Psychology.

Meyer, J.W. & Rowan, B. (1977), "Institutionalized Organizations: Formal Structure as Myth and Ceremony", *American Journal of Sociology*, Vol.83, pp.340-63.

Meyer, M.W. & Zucker, L.G. (1989), *Permanently Failing Organizations*, Newbury Park: Sage.

Miles, R.E. & Snow, C.C. (1986), "Organizations: New Concepts for New Forms", *California Management Review*, Vol.28, pp.62-73.

Miller, D. & Friesen, P.H. (1982), "Innovation in Conservative and Entrepreneurial Firms: Two Models of Strategic Momentum", *Strategic Management Journal*, Vol.3, pp.1-25.

Mintzberg, H. (1979), *The Structuring of Organizations*, Englewood Cliffs: Prentice-Hall.

Mintzberg, H. (1983), *Structures in Five*, Englewood Cliffs: Prentice-Hall.

Morgan, G. (1986), *Images of Organization*, Bristol: Sage.

Morgan, G. & Smircich, L. (1980), "The Case for Qualitative Research", *Academy of Management Review*, Vol.5, pp.491-500.

Mowery, D.C. (1989), "Collaborative ventures between U.S. and foreign manufacturing firms", *Research Policy*, Vol.18, pp.19-32.

Mowery, D.C. & Rosenberg, N. (1979), "The influence of market demand upon innovation: a critical review of some recent empirical studies, *Research Policy*, Vol.8, pp.103-53. Reprinted in Rosenberg (1982, pp.193-241).

Mowshowitz, A. (1990), "On managing technological change", *Technovation*, Vol.9, pp.623-33.

Møltoft, J. et al. (1991), *Kvalitetsstyring og måleteknik* (Quality control and measurement), København: Industriens Forlag.

Nelson, R.R. (1987), *Understanding Technical Change as an Evolutionary Process*, Amsterdam: Elsevier.

Nelson, R.R. (1995), "Recent Evolutionary Theorizing About Economic Change", *Journal of Economic Literature*, Vol.33, March, pp.48-90.

Nelson, R.R. & Winter, S.G. (1974), "Neoclassical vs. Evolutionary Theories of Economic Growth: Critique and Prospectus", *Economic Journal*, Vol.84, pp.886-905.

Nelson, R.R. & Winter, S.G. (1977), "In search of a useful theory of innovation", *Research Policy*, Vol.6, pp.36-76.

Nelson, R.R. & Winter, S.G. (1982), *An Evolutionary Theory of Economic Change*, Cambridge, Mass.: The Belknap Press of Harvard University Press.

Nelson, R.R. & Winter, S.G. (1982a), "The Schumpeterian trade-off revisited", *American Economic Review*, Vol.72, pp.114-33.

Neuve Eglise, M.J. (1986), "CADCAM in France", *CADCAM '86 Newsletter*, April, Birmingham: Coopers & Lybrand.

Northcott, J. & Walling, A. (1988), *The Impact of Microelectronics. Diffusion, Benefits and Problems in British Industry*, Longmead: Blackmore Press.

Nyholm, J. et al. (1994), *Informationsteknologi, produktivitet og beskæftigelse* (Information technology, productivity and employment), København: Kommisionen om Fremtidens Beskæftigelses- og Erhvervsmuligheder.

Otley, D. (1988), "The Contingency Theory of Organisational Control", pp.86-107 in Thompson, S. & Wright, M. (eds.), *Internal Organisation, Efficiency and Profit*, Oxford: Philip Allan.

Pavitt, K. (1984), "Sectoral patterns of technical change: Towards a taxonomy and a theory", *Research Policy*, Vol.13, pp.343-73.

Pavitt, K. & Patel, P. (1988), "The International Distribution and Determinants of Technological Activities", *Oxford Review of Economic Policy*, No.4, Vol.4.

Pay, D. de (1990), *Information Management of Innovations. An International Comparison*, paper presented at the workshop on "The Organisation of Work and Technology: Implications for International Competitiveness", European Institute of Advanced Studies in Management, Bruxelles, May 31-June 1.

Pedler, M., Burgoyne, J. & Boydell, T. (1991), *The Learning Company. A strategy for sustainable development*, London: McGraw-Hill.

Pelz, D.C. & Munson, F.C. (1982), "Originality level and the innovating process in organizations", *Human Systems Management*, Vol.3, pp.173-87.

Perrow, C. (1967), "A Framework for the Comparative Analysis of Organizations", *American Sociological Review*, Vol.32, pp.194-208.

Perez, C. (1983), "Structural Change and the Assimilation of New Technologies in the Economic and Social System", *Futures*, Vol.15, pp.357-75.

Perez, C. (1989), *Equipment, Services and Organizational Change: Three Moving Frontiers for Telecommunications Managers to Handle*, Keynotes Speech at the Canto Conference, Santo Domingo, June 4-9.

Perez, C. & Soete, L. (1988), "Catching up in technology: entry barriers and windows of opportunity", pp.458-79 in Dosi et al. (1992).

Pettigrew, A.M. (1979), "On Studying Organizational Cultures", *Administrative Science Quarterly*, Vol.24, pp.570-81.

Pfeffer, J. (1982), *Organizations and Organization Theory*, Cambridge, Mass.: Ballinger.

Pfeffer, J. & Salancik (1978), *The External Control of Organizations: A Resource Dependence Perspective*, New York: Harper & Row.

Piore, M.J. & Sabel, C.A. (1984), *The Second Industrial Divide: Possibilities for Prosperity*, New York: Basic Books.

Polanyi, M. (1967), *The Tacit Dimension*, London: Routledge & Kegan Paul.

Porter, M. (1980), *Competitive Strategy - Techniques for Analyzing Industries and Competitors*, New York: The Free Press.

Porter, M. (1985), *Competitive Advantage - Creating and Sustaining Superior Performance*, New York: The Free Press.

Revans, R. (1983), "Action Learning: its origins and nature", pp.9-21 in Pedler, M. (ed.), *Action Learning in Practice*, Aldershot: Gower.

Robinson, J. (1933), *The Economics of Perfect Competition*, 2nd Edition, Edinburgh: MacMillan 1969.

Rogers, E.M. (1983), *The Diffusion of Innovations*, New York: The Free Press.

Roos, D. (1991), "The Importance of Organizational Structure and Production System Design in Deployment of New Technology", pp.105-17 in OECD, *Technology and Productivity. The Challenge for Economic Policy*, Paris: OECD.

Rosenberg, N. (1982), *Inside the black box*, New York: Cambridge University Press.

Rosenbloom, R.S. & Cusumano, M.A. (1987), "Technological Pioneering: The Birth of the VCR Industry", *California Management Review*, Vol.29, pp.51-76.

Rothwell, R., Freeman, C., Horseley, A., Jervis, V.T.P., Robertson, A.B. & Townsend, J. (1974), "SAPPHO updated - project SAPPHO phase II", *Research Policy*, Vol.3, pp.258-91.

Sahal, D. (1985), "Technological guideposts and innovation avenues", *Research Policy*, Vol.14, pp.61-82.

Schein, E.H. (1985), *Organizational Culture and Leadership*, San Francisco: Jossey-Bass.

Schein, E.H. (1988), *Organizational Psychology*, 3rd Edition, Englewood Cliffs: Prentice-Hall.

Schoonhoven, C.B. (1981), "Problems with Contingency Theory: Testing Assumptions Hidden within the Language of Contingency Theory", *Administrative Science Quarterly*, Vol.26, pp.349-77.

Scott, W.R. (1992), *Organizations. Rational, Natural and Open Systems*, Englewood Cliffs: Prentice-Hall.

Shimada, "'Humanware' Technology and Industrial Relations", pp.459-70 in OECD, *Technology and Productivity. The Challenge for Economic Policy*, Paris: OECD.

Simon, H.A. (1957), *Administrative Behavior*, New York: The Free Press.

Simon, H.A. (1976), "From Substantive to Procedural Rationality", pp.129-48 in Latsis, S.J. (ed.), *Method and Appraisal in Economics*, Cambridge, Mass.: Cambridge University Press.

Simon, H.A. (1978), "Rationality as Process and as Product of Thought", *American Economic Review*, Vol.68, pp.1-16.

Simon, H.A. (1983), *Reason in Human Affairs*, Oxford: Basil Blackwell.

Solow, R.M. (1957), "Technical Change and the Aggregate Production Function", *Review of Economics and Statistics*, Vol.39, pp.312-20.

Skyum, O. & Dahlgaard, J.J. (1988), *Kvalitet - Danmarks fremtid* (Quality - the future of Denmark), Viby: Centrum.

Statens Industriverk (1988), *Mikroelektronikanvändingen i svensk industry* (The diffusion of microelectronics in the Swedish industry), Stockholm: Gotab.

Stiglitz, J.E. (1987), "Learning to learn, localized learning and technological progress", pp.125-53 in Dasgupta, P. & Stoneman, P. (eds.), *Economic Policy and Technological Performance*, Cambridge, Mass.: Cambridge University Press.

Stufflebaum, D. (1967), "The Use and Abuse of Evaluation in Title III", *Theory into Practice*, Vol.6, pp.126-33.

Sullivan, L.P. (1988), "The seven stages in company-wide quality control", pp.11-19 in Chase, R.L. (ed.), *Total Quality Management*, Exeter: IFS Publications & Springer Verlag.

Sundqvist, U. et al. (1988), *New Technologies in the 1990s*, Paris: OECD.

Tayeb, M.H. (1988), *Organizations and National Culture*, London: Sage.

Teece, D.J. (1986), "Profiting from Technological Innovation: Implications for Integration, Collaboration, Licensing and Public Policy", *Research Policy*, Vol.15, pp.285-305.

Teece, D.J. (1988), "Technological change and the nature of the firm", pp.256-81 in Dosi et al. (1988).

Thompson, J.D. (1967), *Organizations in Action*, New York: McGraw-Hill.

Townsend, J., Henwood, F., Thomas, G., Pavitt, K. & Wyatt, S. (1981), *Innovations in Britain Since 1945*, SPRU Occasionnal Paper Series No.16, University of Sussex.

Tushman, M. & Nadler, D. (1986), "Organizing for Innovation", *California Management Review*, Vol.28, pp.74-92.

Tyre, M.J. (1991), "Managing the introduction of new process technology: International Differences in a multi-plant network", *Research Policy*, Vol.20, pp.57-76.

Urabe, K. (1988), "Innovation and the Japanese Management System", pp.3-25 in Urabe, K., Child, J. & Kagono, T. (eds.), *Innovation and Management: International Comparisons*, Berlin: Walter de Gruyter.

Van de Ven, A.H., Delbecq, A.L., & Koening, R. (1976), "Determinants of Coordination Modes within Organizations", *American Sociological Review*, Vol.41, pp.332-38.

Van de Ven, A.H., Angle, H.L. & Poole, M.S. (1989), eds., *Research on the Management of Innovation: The Minnesota Studies*, New York: Harper & Row.

Walker, W. (1993), "National Innovation Systems: Britain", pp.158-91 in Nelson, R.R. (ed.), *National Innovation Systems*, New York: Oxford University Press.

Whitaker, A. (1992), "The Transformation in Work: Post-Fordism Revisited", pp.184-206 in Reed, M. & Hughes, M. (eds.), *Rethinking Organization*, London: Sage.

Whittaker, D.H. (1990), *Managing innovation. A study of British and Japanese factories*, Cambridge: Cambridge University Press.

Whittaker, D.H. (1990a), *New Technology and the Organization of Work: British and Japanese Factories*, paper presented at the Workshop on the Organization of Work and Technology: Implications for International Competitiveness, European Institute of Advanced Management Studies (EIASM), Brussels, May 31-June 1.

Williamson, O.E. (1964), *The Economics of Discretionary Behavior: Managerial Objectives in a Theory of the Firm*, Englewood Cliffs: Prentice-Hall.

Williamson, O.E. (1975), *Markets and Hierarchies: Analysis and Antitrust Implications*, New York: The Free Press.

Witt, U. (1992), *Evolutionary Economics: An Interpretative Survey*, paper presented at the International J.A. Schumpeter Society Conference, Kyoto, Japan, August 19-22.

Wolffberg, I. & Saabye, M. (1988), *Kvalitetscirkler på dansk* (Quality Circles in Danish), København: Teknologisk Instituts Forlag.

Woodward, J. (1958), *Management and Technology*, London: HMSO.

Woodward, J. (1965), *Industrial Organization: Theory and Practice*, London: Oxford University Press.

Woodward, J. (1970), *Industrial Organization: Behaviour and Control*, London: Oxford University Press.

Zairi, M. (1992), *Advanced Manufacturing Technology*, Malta: Sigma Press.

Zaltman, G., Duncan, R. & Holbek, J. (1973), *Innovations and Organizations*, New York: John Wiley & Sons.

Zucker, L.G. (1977), "The role of institutionalization in cultural persistence", *American Sociological Review*, Vol.42, pp.726-43.

Zaltman, G., Duncan, R. & Holbek, J. (1973), *Innovations and Organizations*, New York: John Wiley & Sons.

Zucker, L.G. (1977), "The role of institutionalization in cultural persistence", *American Sociological Review*, Vol.42, pp.726-43.

Allan Næs Gjerding   Assistant Professor at the Department of Development and Planning, Aalborg University. Previously a Research Fellow at the Department of Business Studies, Aalborg University 1987-93, and Project Manager at Nordjyllands Trafikselskab (the North Jutland Company of Public Transportation) 1993-96. Associated with Lisberg Management 1993-95. Member of the IKE Group at Aalborg University since 1987. At present associated with the International Business Economics at the Centre for International Studies at Aalborg University, and the Danish Research Unit for Industrial Analysis at Aalborg University and Copenhagen Business School.

ISBN 87-7307-526-4